off-site and modern methods of timber construction

a sustainable approach

Robert Hairstans

Stephen,
Thanks for all your help with the illustrations. Champion!

TRADA Technology
TRADA Technology Ltd
Chiltern House
Stocking Lane
Hughenden Valley
High Wycombe
Buckinghamshire HP14 4ND

tel: +44 (0)1494 569600
fax: +44 (0)1494 565487
email: information@trada.co.uk
website: www.trada.co.uk

ISBN 978-1-900510-73-8

Published in 2010 by TRADA Technology Ltd.

Main cover photograph:
Building with standardised components © Lucas Lang

Inset cover photographs, left to right:
Timber panel assembly line © John Meiling
Crane transferring volumetric module © A Falk, Luleå University of Technology
Installation of structural insulated panel roof system © Donaldson Timber Engineering
An entire roof truss system craned into position.

off-site and modern methods
of timber construction

Off-site and modern methods of timber construction addresses a current lack of knowledge on the use of timber in off-site construction. After reviewing the history of timber and off-site construction around the world, the author examines production concepts that improve the efficiency of timber construction. He explains key things to consider when designing buildings intended to be built off-site and reviews the many engineered timber products that facilitate a modern approach to timber construction.

Dr Robert Hairstans has extensive knowledge of the housing market and of lean production systems. Dr Hairstans is Head of the Centre of Off-site Construction and Innovative Structures at Edinburgh Napier University. He was a principal contributor to the Scottish Building Standards Agency Small Buildings Guide and an integral member of the IStructE's UK constitution for the Design of Timber Building Structures to Eurocode 5.

A series of **building case studies**, which includes projects 'built off-site', is available on the TRADA website, www.trada.co.uk. An overview of each project is available to users who register on the site; TRADA members are able to access and download the full case studies. Each case study includes a specification checklist and an aide-mémoire for designers.

TRADA Technology is a world renowned independent consultancy, research and information company, with over 75 years' experience in providing technically-based solutions to help clients make informed business decisions and add value to their operations, products and services. Although the company's core skills relate to timber, it also works with other building materials and more generally throughout the construction sector.

TRADA Technology is an independent consultancy company providing a wide range of commercial and training services to the timber and construction industries. Prior to 1994 it was wholly owned by TRADA, the Timber Research and Development Association. It is now a member of BM TRADA Group of companies and is TRADA's appointed provider for its Research and Information programmes and for the administration of its membership services.

For online technical information, details of consultancy services, training and TRADA membership visit www.trada.co.uk or telephone 01494 569600.

Also within the BM TRADA Group is **BM TRADA Certification,** a leading multi-sector certification body accredited by UKAS (United Kingdom Accreditation Service). BM TRADA's potfolio of services includes certification to quality, environmental, information security, and health and safety management systems. Its certification services include chain of custody – FSC, PEFC, and Forest Product schemes.

For certification and chain of custody information visit www.bmtrada.com or telephone 01494 569700.

TRADA, the Timber Research and Development Association, is a not-for-profit, membership-based organisation delivering key services to members in support of its two main aims of 'Building markets for timber' and 'Increasing specification'. Membership encompasses companies and individuals across the entire timber supply/use chain, from foresters and sawmillers, through merchants and manufacturers, to architects, engineers and specifiers.

For further information and details of membership visit www.trada.co.uk or telephone 01494 569603.

Contents

1 Introduction

In many respects timber is the ideal construction material. It is strong in both tension and compression, has a high strength-to-weight ratio and can be relatively easily worked. If appropriately sourced, timber can be considered truly sustainable and, with good silvicultural practices, trees can also improve land quality and soil fertility.

Primarily, however, the human race has employed timber for building because it is available in most habitable regions of the world. The climate and conditions of the region have had a bearing on the nature and species of tree growing which subsequently influenced the end form of construction. These forms of timber construction have evolved over time due to mechanisation and the invention of new timber products and connection methods. Modern procurement methods and building regulations have also influenced building forms.

Off-site construction is the manufacture and pre-assembly of components, elements or modules before installation into their final location. Urbanisation and colonial expansion prompted the need for efficiency and quality of product, resulting in the first major exploitation of off-site techniques, although there is also evidence of off-site concepts being used by medieval carpenters. With the evolution of construction components and techniques, varying forms of off-site construction have also progressed to become Modern Methods of Construction (MMC).

Chapter 2 gives an overview of the products (including panelised units, volumetric and hybrid techniques) and processes. It compares the UK and global timber construction markets with respect to system evolution, procurement methods and specifics of design, detailing and products used. This chapter considers the drivers for the use of off-site construction along with the concept of applying lean theory in order to continuously improve the off-site production and on-site construction environments.

MMC are about better products and processes. They aim to improve business efficiency, quality, customer satisfaction, environmental performance, sustainability and the predictability of delivery timescales. MMC are, therefore, more broadly based than a particular focus on product. They engage people to seek improvement in the delivery and performance of construction through better processes. *Chapter 2* includes a robust overview of research and practice in this area as well as the findings of the various Government task forces such as Barker and Latham.

The procurement process in off-site MMC is important. Private and social contracts have different demands, as do one-off build projects compared to, for example, volume house builders. Each sector makes different demands on the supply chain and logistics, and the relative level of standardisation and mass customisation that can be achieved varies. The business model is therefore important for success, particularly for striking the correct balance between the level of preassembly and construction activities. Another key driver to off-site MMC is customer or client satisfaction of the quality of both product and service. Partnering facilitates this and is therefore presented as a key concept to the success of off-site MMC.

Chapter 3 introduces the principles of production and shows how these can be selectively applied to construction. This chapter summarises lessons learned in production (manufacturing) and the tools and techniques, borrowed from production, that are embodied in off-site MMC. Case studies demonstrate pioneer projects that have successfully applied production principles in timber construction.

The environmental performance and sustainability of building systems is becoming more critical. *Chapter 4* examines the impacts that new legislation and consumer awareness of the energy performance of buildings are making on timber forms of construction, and how researchers and developers are addressing these challenges. This chapter includes an overview of UK sustainability policy and regulations, together with case studies that show how timber construction is responding to The Code for Sustainable Homes and the Passive House concept. The use of timber as a sustainable form of construction is assessed and the concept of Life Cycle Analysis is presented including information on ensuring timber is sourced from sustainably-managed forests.

Chapter 5 reviews the design of timber buildings and the additional consideration required when applying off-site MMC with particular focus on main issues such as differential settlement, structural stability and disproportionate collapse. This chapter also presents Eurocode, which has now superseded British Standards, and provides the rationale as to why it will prove advantageous to use especially when considering off-site MMC.

Engineered timber products are key components in off-site and modern methods of timber construction. *Chapter 6* begins with an overview of the structure of timber. It summarises the properties and species of timber available, how the properties affect the end application and what its limitations are. It goes on to show how timber is being re-engineered to overcome the limitations of its natural form, as well as how innovations are improving timber products for use in MMC. Modern truss systems, timber composites and engineered products are all explained, together with their methods of fabrication and applications.

This book draws on a multitude of sources including regulatory documents, research outputs and case studies in order to describe how off-site timber construction has evolved to become a modern method. From this, the principles of off-site MMC are explained together with the ways in which these principles can be used to deliver a more cohesive and socially, economically and environmentally sustainable industry by applying improved business models. These include partnering arrangements, robust product procurement strategies, material optimisation and system efficiencies that, in combination, enable the industry to continually evolve and to innovate through applied research and development. And when these concepts are embraced, timber can readily be seen to be the ideal material for Modern Methods of Construction.

2 Off-site modern methods of construction

2.1 Off-site construction: a brief history

Off-site construction is primarily based on two key principles: efficiency and quality. It is defined as the manufacture and pre-assembly of components, elements or modules before installation into their final location. The concept of off-site construction is to allow parts of the building to be manufactured in an environment suited for effective production, where advanced equipment can be used to increase productivity. Furthermore, off-site construction improves working conditions by taking construction processes to a clean, controlled environment unaffected by climatic conditions. The more parts of buildings completed off-site, the less needs to be done on-site.

There is evidence of off-site construction throughout history. In the UK the concept was utilised in the medieval period where a degree of 'off-site' construction was used as a quality assurance procedure. Carpenters would often fabricate the frames of buildings in their yards, the joints created and the frames assembled to ensure that all the elements fitted accurately together. The frames were then dismantled (after all the joints were carefully marked with 'carpenter's marks'), transported to site and reassembled.

It took until the beginning of the 19th Century before the advantages of off-site construction were first truly exploited. This development can be ascribed to urbanisation and colonial expansion that together necessitated buildings to be produced, marketed, packaged and dispatched in large quantity to all corners of the earth. Constructed in England and subsequently shipped to Australia, timber housing sections were used in 1830 to provide a "*comfortable dwelling that can be erected in a few hours after landing, with windows, glazed doors, and locks, bolts and the whole painted in a good and secure manner*" (1). This off-site concept was pioneered by London carpenter and builder, John Manning. Prior to Manning there were examples of off-site construction in the form of elementary timber-framed and sheathed huts and sheds as well as isolated examples of ambitious, specially conceived, fabricated, and erected houses for wealthy clients. However, it was the production of the Manning cottage (*Figure 2.1*) which embraced the key off-site concepts. The Manning cottage was standardised for efficiency ("*every part of it being made of exactly the same dimensions: that is, all the panels, posts and plates, being respectively of the same length, breadth and thickness, no mistake or loss of time can occur in putting them together*" (1)). And it was designed to accommodate logistical requirements (mobility and ease of transport) whilst being simple to construct due to a lack of skills ("*completely fabricated in the carpenter's shop and required only the foundations to be built and the components assembled*" (1)).

Figure 2.1 John Manning's portable colonial cottage
Photo: Herbert (1)

The concepts of on-site inefficiency were identified in 1917 by a government committee which discovered that bricklaying (materials and labour) accounted for 31% of the total construction cost with joinery and carpentry accounting for 26%. The purpose of removing on-site activities to a factory environment was to increase productivity through the implementation of efficient manufacturing procedures and timber as a material lent itself far more readily to this process than brick.

The off-site concept even in the early years was not limited to housing, with firms receiving British Government grants to produce churches, chapels and schools. Britain therefore pioneered the concepts of off-site construction but America - due to the fact that timber was a more prevalent construction material both in terms of skill base and availability - was better equipped to meet the demands of the global market.

Market demand is a common driver for the use of off-site construction. In post Second World War Britain there was unprecedented need for public housing. To meet the housing requirement, mass production was identified as the solution and the concept of system building was endorsed whereby housing would be mass produced in a manner reflecting the production of cars and aeroplanes. However, although this "revolution to building methods" was implemented with good intention (to fulfil the housing need with better quality whilst improving working conditions) it did not fully work as intended but only acted as a step in the evolutionary process [(2)]. System build failed mainly as a result of being production-focused and non-customer orientated, resulting in poor public perception and architectural apathy. Clearly any future industrialised housing approach would need a customer focus and processes for integrating design, production and information management.

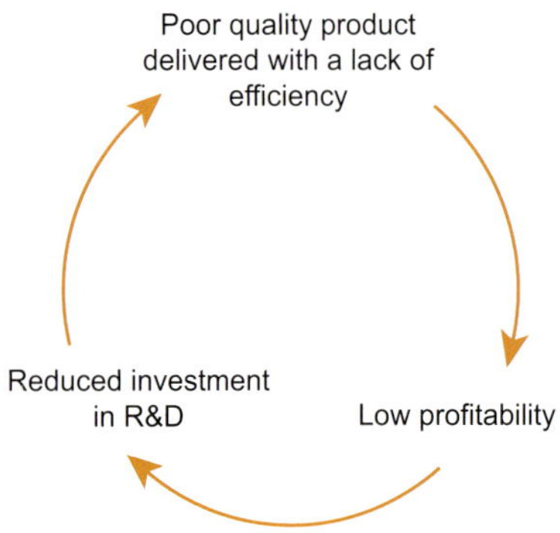

Figure 2.2 Degenerative cycle

More recently, in 1998, as a result of concerns about the state of the industry a report was commissioned by the Construction Task Force to the Deputy Prime Minister [(3)] on the scope for improving the quality and efficiency of UK construction. The overall concern was that a degenerative cycle was taking place: poor quality product was being delivered with a lack of efficiency, resulting in low profitability and subsequent reduced investment in research and development.

The challenge set by the Task Force was not for the construction industry to improve its current methods, but for the industry and the Government to join with major clients and do it entirely differently. In essence the proposal was to "rethink" construction.

The UK government acknowledged that construction must be "re-thought" and that the underlying reasons for the lack of responsiveness and supply should be reviewed. Demand for housing in the UK is increasing over time driven primarily by demographic trends and rising incomes. Yet in 2001 the construction of new houses in the UK fell to its lowest level since the Second World War and, between 1992 and 2002, output of new homes was 12.5% lower than for the previous decade [(3)]. The economic down turn experienced in 2008 and 2009 resulted in a drastic reduction of 57% in housing starts across the regions of Britain (*Table 2.1*). However, the UK Government still requires 3 million new homes to be produced by 2016 and, as a result, a step change is needed in the ability of the construction industry to meet demand.

2.2 Modern Methods of Construction

In 2004 the Barker 33 Cross-Industry Group was established and involved stakeholders from 50 separate organisations across the construction sector. Its purpose was to examine the barriers to the greater use of Modern Methods of Construction (MMC).

Table 2.1 UK House Building Starts (NHBC statistic)

Region	Jan to Mar 2009	Jan to Mar 2008	Percentage Increase / (-) Decrease
England - Regions			
North East	605	1,253	-52%
North West	1,105	2,578	-57%
Merseyside	385	966	-60%
Yorkshire and the Humber	1,027	2,567	-60%
West Midlands	929	2,898	-68%
East Midlands	1,219	3,240	-62%
Eastern	2,570	4,157	-38%
South West	1,745	3,088	-43%
Greater London	1,304	4,427	-71%
South East	2,501	6,349	-61%
Totals for England	**13,390**	**31,523**	**-58%**
Scotland - Councils	1,446	4,004	-64%
Wales - Unitary Authorities	729	1,425	-49%
Northern Ireland - Counties	665	736	-10%
Isle of Man	2	151	-99%
Totals for UK	**16,232**	**37,839**	-57%

According to the group (4),

> *"Modern Methods of Construction are about better products and processes. They aim to improve business efficiency, quality, customer satisfaction, environmental performance, sustainability and the predictability of delivery timescales. Modern Methods of Construction are, therefore, more broadly based than a particular focus on product. They engage people to seek improvement, through better processes, in the delivery and performance of construction."*

The findings of the review procedures resulted in further investment in MMC. Initially the Sustainable Communities Plan, launched in 2003, aimed to encourage the use of MMC to produce the quantity and quality of houses needed to meet the 2016 target of 3 million new households (5). Subsequent to this in 2005 the Office of the Deputy Prime Minister (ODPM) and the Housing Corporation spent £1.1 billion building affordable housing using MMC, including £0.5 billion using off-site manufactured products (6).

MMC is defined by the ODPM (now Communities and Local Government) as being *"a process to produce more, better quality homes in less time"*, MMC therefore requires to be both efficient and effective. Within the ODPM definition of MMC four product sectors have been derived:

- *Panelised units*: produced in a factory and assembled on-site to produce a three dimensional structure. The spectrum of these ranges from open panels that consist of a skeletal framework only to advanced panel systems that can incorporate lining material, insulation services, windows, doors, internal wall finishes and external cladding.

- *Volumetric construction*: three dimensional modular units produced in factory conditions prior to transport to site.
- *Hybrid techniques*: panelised and volumetric approaches are combined. An example of this is the use of volumetric units (also referred to as pods) used for highly serviced and more repeatable areas such as kitchens and bathrooms, with the remainder of the dwelling or building constructed using panels.
- *Other*: constructions that may use floor or roof cassettes, pre-cast concrete foundation assemblies, preformed wiring looms, and mechanical engineering composites.

Figure 2.3 shows a timber example of each product sector.

It is worth noting that adopting a particular product sector does not necessarily mean that the overall ethos of MMC has been adopted. The use of product must be combined with mechanisms for ensuring business efficiency, product quality, predictability, workforce engagement and importantly customer satisfaction.

Timber platform frame open panel construction

Volumetric massive timber unit [7]

Fabrication of serviced volumetric OSB4 modular bathroom units (Panaloc System)

Floor cassette

Figure 2.3 Different forms of timber off-site MMC

Photo of Volumetric massive timber unit: A Falk, Luleå University of Technology

It is important to note that MMC do not necessarily have to be conducted off-site. On-site efficiency and quality can be improved by employing improved management techniques. Removing on-site construction activities to an off-site controlled environment therefore allows the conversion of construction assignments to a production flow process which in turn facilitates the application of Scientific Management Theory [8]. Scientific Management Theory stipulates that management can maximise the productivity of each worker by prescribing, as a result of exact measurement and detailed study, the most efficient way of carrying out operations.

Moving processes off-site depends on the specifics of the project or market to be supplied (ie private, public, one-off, mass production, location etc) and the relative demands it places on the supply chain and logistical activities. The supply chain can be considered to be in two main parts:

- Preassembly factories: purchasing, material handling, supplier involvement, prefab production, transportation and supply patterns etc
- Construction site: orientation of trades, material delivery and storage, sub or final assembly etc.

And with respect to the above the decision making process should be holistic when considering the level of off-site construction to be used (*Figure 2.4*).

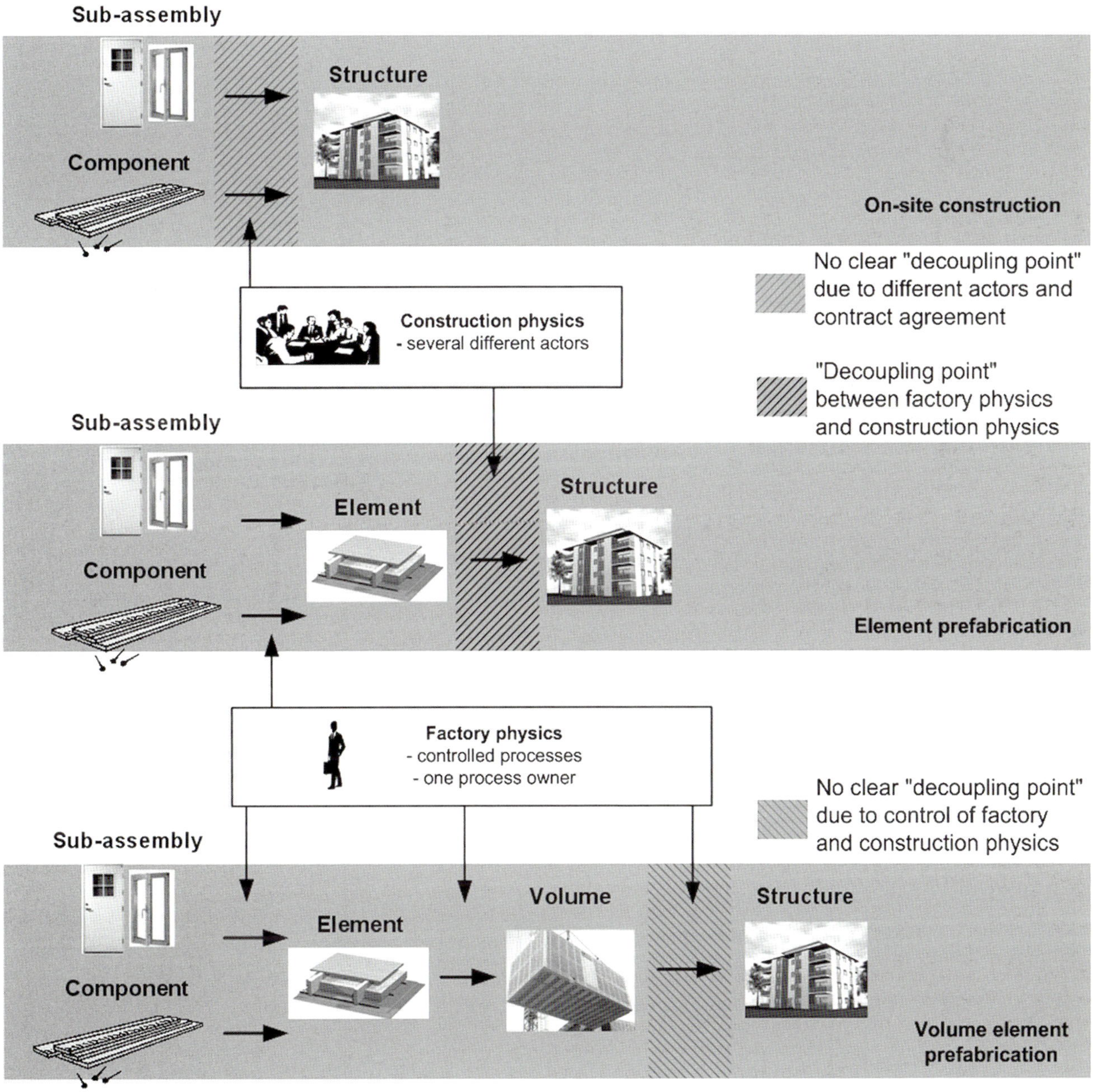

Figure 2.4 Level of off-site construction
After Sandberg et al [9]

In accordance with the recommendations of the ODPM, achieving the balance between preassembly and construction, activities must be implemented in a manner that meets the requirements of the MMC ethos. In order to achieve quality and efficiency, the waste, lead time, variability and complexity must be reduced whilst maintaining the flexibility required. Accordingly, the more activities that are moved upstream from the construction site to factories where pre-assembly is carried out, the higher the demands on management of the supply chain and logistics. Further, the higher the level of automation required, the higher the level of capital investment, which produces difficulties in terms of achieving economies of scales. Other barriers that restrict the take up of off-site activities are the need to freeze design early and, as a result of recent history, poor perception by consumers. Production concepts from the manufacturing industry require adaptation to properly match the construction industry's needs and demands, highlighting the importance of finding a balance between project and process focus [10] [11].

The underlying economics of off-site MMC, and modular construction in particular, are quite complex and require an efficient business model to be effective. Economics of scale and repeatability are important but must be balanced with the required level of flexibility in order to satisfy the demands of the target market. MMC, other than open panel techniques, continue to be slightly more expensive than more established techniques and highly documented off-site MMC techniques (complete modular building, bathroom and toilet pods and flat pack, kitchen flat pack, off-site plant room and closed wall panels) are actually only applied to a very limited extent in housing. However, higher buildings do favour volumetric MMC because construction costs rise faster with brick and block.

In terms of product quality, MMC can deliver at least as good a quality as more established building techniques, providing that they are adequately specified. Although off-site manufacture may not guarantee enhanced durability greater than traditional construction methods, *"factory production should reduce the risk of non-conformities, related premature failures and consequent repairs which may be associated with on-site assembly"* [6]. However, MMC is not just about delivering a commodity; it is also about delivering a higher level of service and customer satisfaction. Applying MMC correctly will also result in project time savings, if the process plans are tailored to match the method of construction. Mitigating the major MMC project risks requires process discipline, good coordination and a culture that will not accept late changes. The main risks relative to MMC are:

- *Late design changes*: difficult to absorb because factory work based on the design starts early. Well established partnering arrangements and close collaboration between parties is therefore good practice.
- *Loss of a factory production slot*: normally as a result of poor communication and can result in lengthy delays. An effective and robust communication mechanism is therefore necessary.
- *Building tolerances and accuracy*: foundations in particular require tolerances to be accurate. Products delivered to site have a high degree of dimensional accuracy and, as a result, reciprocal foundations tolerances are required.
- *Supply failure*: for the time saving of MMC to be realised, the elements need to be supplied and erected on time. To ensure this happens, it is important to implement procedures for the management of materials,

products and information flow, also known as supply chain management (SCM) [12].

An important feature of MMC is continuous product and process improvement. Whether this is improved health and safety, meeting new regulatory requirements or better built quality, the concept of MMC requires everyone on the supply chain to work towards a better end result, through a wide range of mechanisms.

2.3 Evolution of timber construction

Based on the definition that a tree is a woody plant growing on a single stem usually to a height of over two metres, approximately 21,000 different species exist across the globe. Although the species are diverse, commercially they are divided into two categories: softwoods and hardwoods. Softwoods are generally evergreen with needle-like leaves (which in biological terms are known as gymnosperms; ie plants bearing naked seeds) and include the spruces and pines. Hardwoods are generally broad-leaved (deciduous) trees that lose their leaves at the end of each growing season and include birch and oak.

Most climatic zones have at least one species that has adapted to the prevailing conditions within that area. Timber is therefore available in most habitable regions of the world and, as a result, the human race has used it extensively in 'construction' to provide a means of shelter and protection. Over the centuries timber construction has evolved substantially with respect to the given resources available (tree type, technology, skills etc) and market (culture, demand etc). *Figures 2.5, 2.6 and 2.7* show how beam and post, timber frame and massive forms of constriction have evolved.

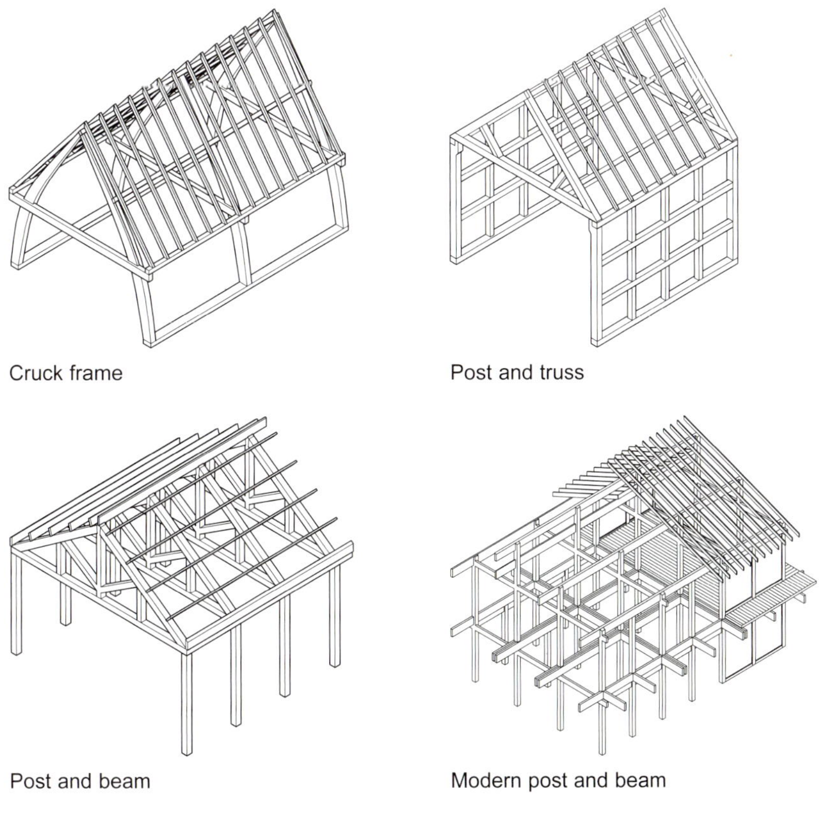

Figure 2.5 Evolution of timber post and beam construction

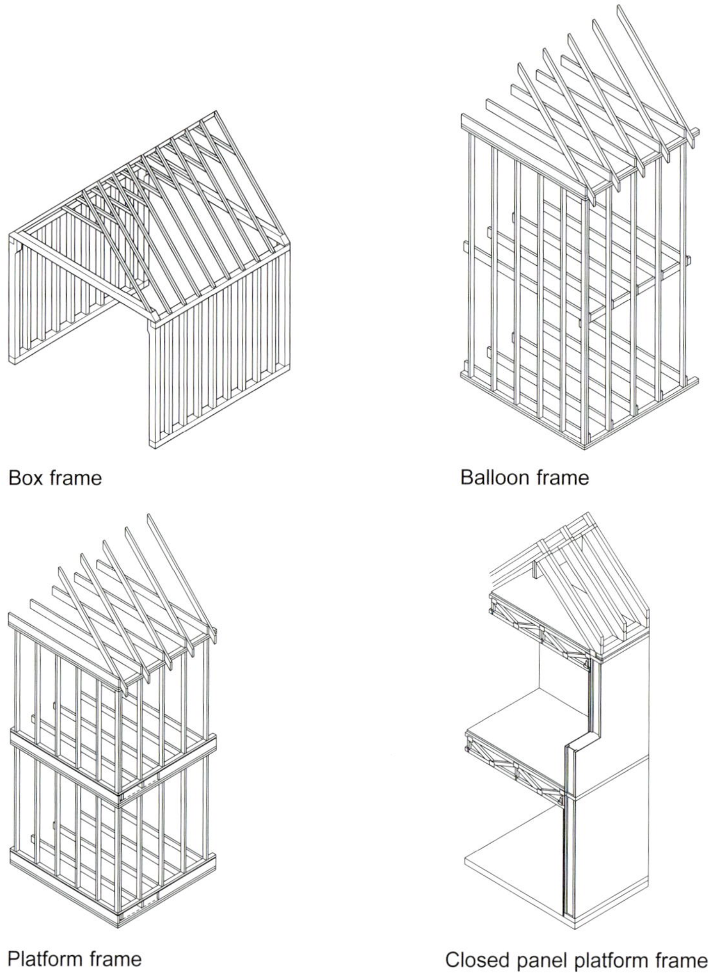

Figure 2.6 Evolution of timber frame construction

2.3.1 Early Timber Structures

Climatic conditions and natural environment largely dictated the form and function of early timber structures. Seismic activity is present in the natural environment of the Far East and, as a result, heavy roof structures supported on sophisticated frame systems that originated in China were

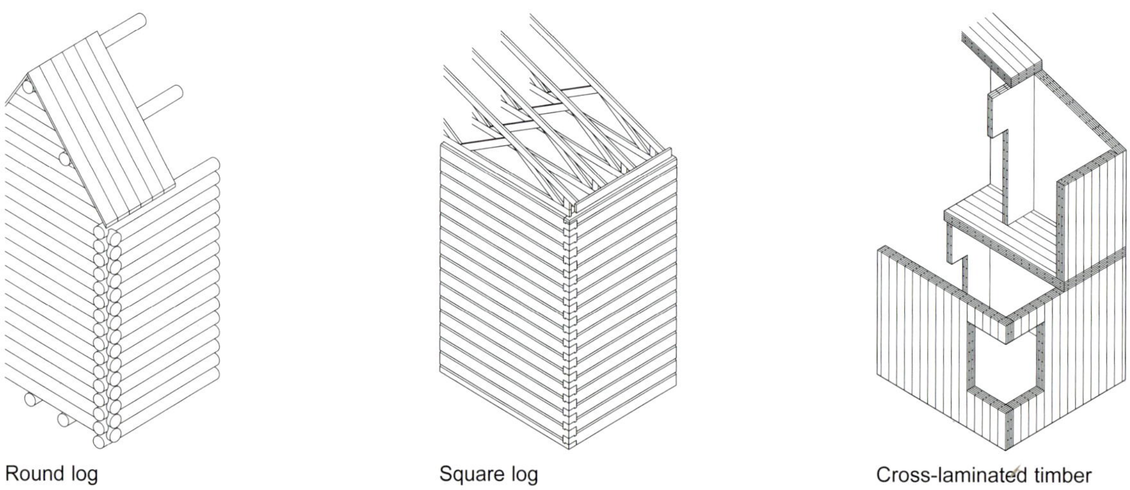

Figure 2.7 Evolution of massive timber construction

favoured. Japan contains the most surviving wooden structures and those that still exist are timber frame systems joined by intricate brackets sets that provide great strength and stability.

Timber construction is not restricted to frame structures, particularly where there is an abundance of resource which is the case in both the Northern and Southern regions. Around the North Pole there is a ring of softwood that encircles the planet and further south, and at lower altitudes, grow a variety of deciduous trees. The abundance of resource in these regions meant there was no scarcity of material to build and the major traditional method was "log construction" (block work).

Block work construction methods are still employed today, again mainly in global regions, where a higher degree of indigenous resource still exists such as the Scandinavian countries (*Figure 2.8*). There are methods of construction that are considered more modern that are employed primarily in areas where there are high levels of indigenous resource and follow a similar concept of solid timber section in the form of massive timber systems. Massive timber systems are fabricated from cross laminations of timber bonded by glue or connected by a timber dowel technique (brettstapel).

Figure 2.8 Examples of modern block work developments in Scandinavia

2.3.2 Traditional forms in Britain

Unlike the softwood timber available in these countries most associated with block work construction, Britain's hardwood timber resource – even in its primeval forests – has never produced the long, uniform and only slightly tapered tree trunks necessary for this method to be employed. As a result, traditional timber construction in Britain, and in other parts of continental Europe, generally took on a different form of which there are three main categories:

- Cruck construction
- Box frame construction
- Post and truss construction.

In one form or another, these forms of construction are still employed today in the form of modern oak frames (*Figure 2.9*) and beam and post systems. Whilst oak frame systems tend to still use traditional timber dowel connection methods, other systems often employ bolted steel connections.

Figure 2.9 Modern cruck frame construction
Photo: The Green Oak Carpentry Company

2.3.3 Germany

The German system is typically post-and-beam, structural insulated panels or a combination of both. The houses are generally considered to be of 'chalet' design but other forms are marketed, and there is considerable variety and flexibility in external features. Typical houses in Germany have open living areas, large areas of glazing, basements and the fitting and final finish are invariably of a high quality, although some supply different levels of finishes to match customer budget. The range of homes supplied by off-site manufacturers serves a wide market from 'starter' homes to luxury housing. The upper end of the market will undertake to build to any plan desired by the customer while the cheaper houses offer less flexibility. However, internal layouts and specification can be changed.

The hybrid nature of these systems gives designers considerable flexibility and the capability to produce closed panels with the inclusion of insulation, services, linings, windows, doors and cladding in the factory although the majority of product leaving the factory is a basic frame with only linings, insulation and service conduits installed.

2.3.4 New World

Timber frame construction as we know it today really began to take shape in the New World during the nineteenth century. In North America the population was rapidly expanding and there was a need to rationalise the way houses were built to satisfy demand. The demand for housing coincided with the development of mechanised saw mills, as these could produce accurate and small sizes of timber, as well as the wire nails that made intricate, hand crafted joints redundant. As a result of this the skeletal frame work was born, not just in North America but also in the other major softwood producing countries around the globe.

The pressure to build rapidly in North America resulted in the invention of 'balloon frame construction', evidence of which can be traced back to methods employed by seventeenth century carpenters in Virginia. Balloon framing is two- or three-storey height timber framed and sheathed wall panels which act as vertical diaphragms and support roofs and floors acting as horizontal diaphragms.

Chicago gained a reputation for inventing the balloon frame and Chicagoan George W. Snow is credited as the man who revolutionised construction practice with its invention in 1832. The city however adopted this reputation due to the fact that its factories produced ready-made houses with balloon frames that were sold to various western cities attempting to meet the needs of their rapidly expanding populations. Platform timber frame is a derivative of the balloon frame and is most commonly used in Canada, the USA and the UK.

Similarities between North America and Australia can be drawn in terms of methods of construction and the reasons for it. In 1851 gold was discovered in Victoria and, in only three years, the population of the colony swelled from 76,000 to 540,000 The bulk of the timber for the required houses was imported from the west coast of America (often Oregon pine) and presumably with it the similar methods of timber frame construction used there.

Under construction   Stone cladding   Timber cladding

Figure 2.10 Timber platform frame - USA

Timber platform frame is still the most predominant method of construction in the USA today (*Figure 2.10*).

2.3.5 Japan

Wood construction accounts for 40-50% of the more than 1.2 million housing starts per year in Japan (JAWIC, 2001). Annually the country builds the second highest number of wood houses in the world, the predominant form of which is post and beam. However, timber platform frame methods of construction, first introduced in 1972 from North America, now account for a growing share of the Japanese market (*Figure 2.11*).

In Japan 80% of the house and land market is segregated with land either purchased or inherited. Land within major population centres is limited, resulting in houses being built within close proximity which restricts available construction methods and necessitates the need for robust fire proofing. From a structural robustness perspective, seismic activity and typhoons need to be taken into consideration during design. Further, the building performance of Japanese housing needs to take into account a wide range of environmental conditions ranging from temperate in the North to sub tropical in the South.

Until relatively recently Japanese house construction was not designed for the long term. Housing quality assurance laws in Japan require homes to come with a minimum guarantee of 10 years. As a result of this the market became accustomed to low lifespan construction (20 – 40 years). This had market impacts with houses being acquired for the land and subsequently demolished for a new build to be constructed (contributing to a high volume market). However, the Japanese market is shifting and in 2006 the Japanese government enacted the Basic Act for Housing, a driver being the need for sustainability. A consequence is that a house should not be consumed by one individual but rather it should be treated as a part of the social infrastructure. The industry has changed from building houses with a potential 30 year lifespan to building long (40 year) and ultra long life (100 year plus) builds. The phrase "200 year housing" was coined as an ultimate objective.

The off-site sector is capable of meeting the modern Japanese housing requirements and is becoming more prevalent due to a number of key issues:

- Consumer demand is for a higher standard of product which off-site construction can address more readily.
- Suppliers of off-site housing have conducted extensive market research to identify consumer requirements tailoring their standard components to these needs.
- Site access and storage restrictions result in the requirement for a systematic and organised approach to construction.

In Japan the production concepts are applied to industrialised housing and, as a result of balancing customisation and standardisation, efficient production processes have been developed. The off-site manufacturers of housing are driven by innovation and use research and development as a marketing tool, engaging directly with the consumer primarily through exhibition parks and promotional centres where recent developments and products are showcased. The major house producers promote their products not just as homes but "living spaces" and "building blocks for communities".

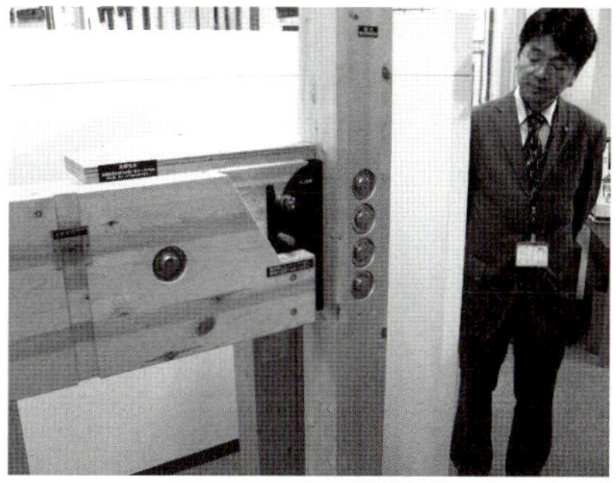

Connection Detail (Sekisui Homes)

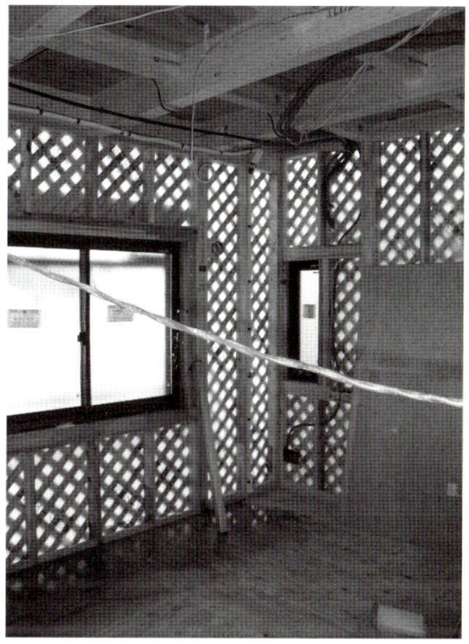

Timber frame construction (Sumitomo Forestry)

Figure 2.11 Timber construction in Japan

2.3.6 Finland – light weight framing

In the 1900s, the concept of lightweight timber frame construction was imported from North America to Finland. In the beginning, take up of this timber frame method was slow, because block work was still the predominant method of construction as logs were readily available and relatively cheap. However, during the 1930s timber frame construction became the dominant construction method in Finland as a result of it being less laborious and a more efficient use of resources. By the 1980s, timber frame construction in Finland had evolved to timber wall with mineral wool insulation and a brick façade [13] (*Figure 2.12*). In many respects this method of timber frame construction resembles UK timber frame wall construction today, although modern timber frame systems in Finland are now mostly clad with timber.

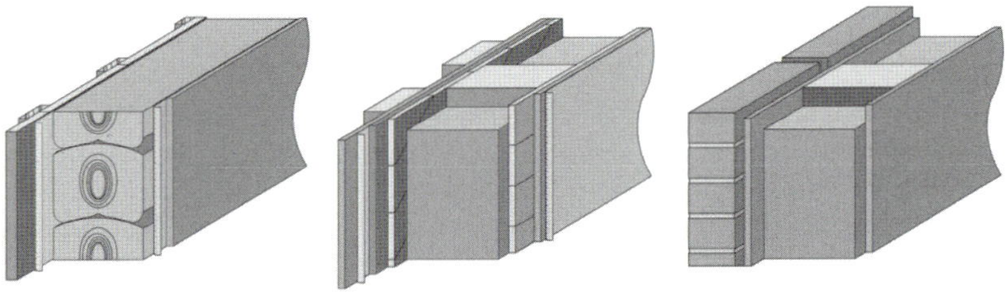

Figure 2.12 Evolution of the timber wall in Finland
After Risto Suikkari

Log wall in the 1930s | Timber wall with saw dust insulation 1950s | Timber wall with brick façade and mineral wool insulation in 1970s and 1980s

2.3.7 Sweden – timber volume elements

In Sweden, off-site timber construction is in a strong position particularly in the private residential market. The traditional procurement process was displaced in the 1970s when, as a result of a demand for single-family housing, direct relationships between customers and the manufacturers were commonly created. With customers going straight to the manufacturer, rather than buying from a developer, houses were optimised to meet end user requirements based on prefabricated standardised elements, resulting in rational and efficient production and lower costs. The Swedish market in single family housing has therefore developed to supply efficiently produced standardised components within a process that has been adapted to customer demands [14] (*Figure 2.13*).

Current off-site timber frame construction accounts for approximately 80% of the Swedish market [15]. Within this market three types can be distinguished: off-site prefabrication of materials and parts, prefabrication of components

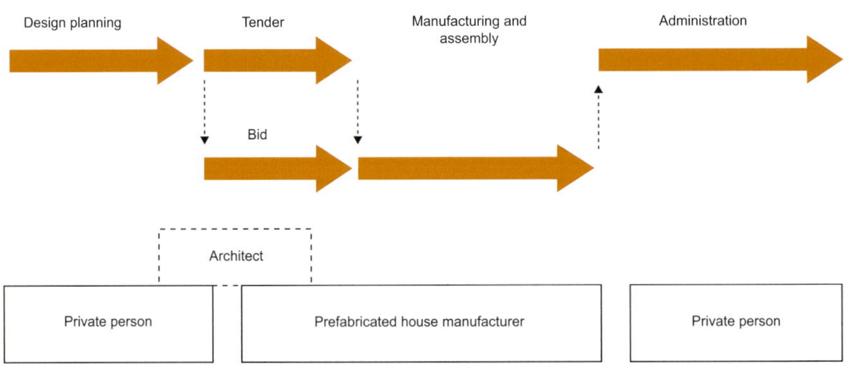

Figure 2.13 Typical single family home procurement process in Sweden
After Sandberg [14]

and subassemblies, and timber volumetric construction or "timber volume element" (TVE) prefabrication (*Figure 2.14*) where 80% of the work is completed off-site in a factory environment (16) (17) (18).

TVE construction in Sweden can be traced back to the 1950s where it was used to produce detached housing or standardised booths and barracks in a repetitive production process. It was subsequently also used for schools and office buildings (15). Currently TVE in Sweden accounts for around 30% of the detached housing market, but somewhat surprisingly much less (approximately 8%) of the multi-storey market. Volumetric construction tends to lend itself more to multi-storey as a result of repetition. However, until 1995 timber construction in Sweden was restricted to low rise construction and there has also been a degree of consumer scepticism in relation to building performance (primarily acoustics) when used for multi-storey construction.

The TVEs manufactured in Sweden are built up of timber based elements and consist of four load bearing walls (exterior walls or volume separating walls), a system of joists, interior roof and a number of partition walls. The TVE manufacturers prefer to have control over the whole project process, from design to complete building. Professionals (architects, engineers, contractors etc) working on projects are either in-house or are engaged through strong partnering arrangements in order that they are well versed in company procedures, manufacturing and working methods. The following manufacturing process takes place on static work stations:

1. The frame of the elements is finished.
2. Insulation and services are installed.
3. Boards are placed on the frame to seal out the elements and subassemblies such as doors and windows are installed.
4. Once completed the elements for one volume are moved to a volume assembly station.
5. The wall elements are placed and fixed on to the floor element and the roof element is placed on top
6. The volume elements are then completed with flooring, finishing, installations, wardrobes, cabinets and white goods are installed as required.
7. Completed volumes are covered with moisture-proof tarpaulins prior to transportation.

Critical to the success of TVE construction is the early design phase where the whole process can be improved through standardisation. Standardisation can be taken to be in conflict with flexibility, however, and informed customisation is only achievable if the customer understands the relative effect on lead time and cost, hence the reason why partnering is important. Nevertheless, TVE layout variation is limited as a result of the volumes being built up on four load-bearing walls and there are also restrictions on what can be achieved in terms of length, width and height due primarily to manufacturability and logistical requirements. In order to achieve Swedish building performance requirements for airborne and impact sound insulation, the floor system needs to be relatively thick. In conjunction with this, volume separating walls also result in doubly thick interior walls. There is therefore an over specification of required material and a need for increased building footprint relative to habitable volume (15).

Elements ready for assembly

Elements assembled into volumes, including internal finishes, kitchen and bathroom furnishings, and services

Installation

Figure 2.14 TVE in Sweden
Photos: John Meilling

2.3.8 Britain today

In the UK there are examples of modern type timber frame houses that date back 100 years. However, it was amendments to the Building Regulations in April 1965 and a call for more houses to be built using less labour that stimulated the use of timber platform frame. For the first time, the regulations limited the amount of energy that could be lost through certain elements of the fabric of the building (expressed as U-values – the amount of heat lost per square metre, for each degree of Celsius of temperature difference inside and outside). Since 1965, growth has remained steady and now accounts for approximately 25% of the UK market. It is envisaged that the UK timber platform frame (*Figure 2.15*) market will continue to grow as a result of its ability to deliver durable and sustainable housing at a fast rate.

Figure 2.15 Standard two storey UK timber frame house

Under construction

Completed house

The majority of timber platform frame construction in the UK is now an integrated process from design to erection. At the design stage the roof and floor systems are designed using the software associated with the specified product. The structural frame work is designed by a structural engineer and drawings of the frame subsequently produced using a CAD-based design package.

The timber frame system has a relatively high degree of compatibility and can be conceived as composite wall and floor units built up from timber framing, panel products, insulation and cladding. The composite units in a timber frame system can be utilised for:

- Transfer of vertical loads
- Stabilisation of dynamic and seismic loads
- Physical separation
- Sound insulation
- Thermal insulation.

The design is required to optimise all the relevant requirements; however, in the UK this is rather difficult given the fragmented nature of industry. External consultant engineers, roof truss suppliers and floor component suppliers are often all involved in the design process. In terms of the design of walls and floors, different aspects can be identified as critical and in order of priority are:

1. Fire resistance
2. Horizontal stabilisation
3. Sound insulation
4. Vertical loading.

The roof trusses are normally fabricated externally by the roof truss manufacturer, and floor cassettes and open wall panels are fabricated off-site by the timber frame manufacturer, in controlled factory environments, and then delivered to site for erection. All the materials are cut from automated cutting lists, produced during the drawing process (CAD/CAM software) by using optimising saws to reduce waste. The fabrication of floor cassettes and wall panels is normally a manual process with production line procedures adopted for efficiency. However, frame makers are becoming more common for the fabrication of walls. As a result of the use of timber composites and engineered products in the manufacturing of wall panels and floor cassettes, product dimensions are not restricted by available stock sizes but are dictated by transportation issues and the erection method.

The standard factory produced structural wall panels (*Figure 2.16*) consist of sawn timber framing material (normally 38 or 45mm wide by 89, 95, 115 or 140mm deep, C16 grade studs at 600mm centres) sheathed with single or double 9 or 11mm OSB (depending on the required structural performance). Stud depth is predominately 140mm thick now as a result of building regulations requiring increase levels insulation to meet the require U-values.

The cassette flooring system normally consists of 240mm deep I-joists spaced at 600mm centres. The flange specification of the joists will be determined by the load span conditions and, in certain design circumstances, deeper I-joists at closer centres may be required.

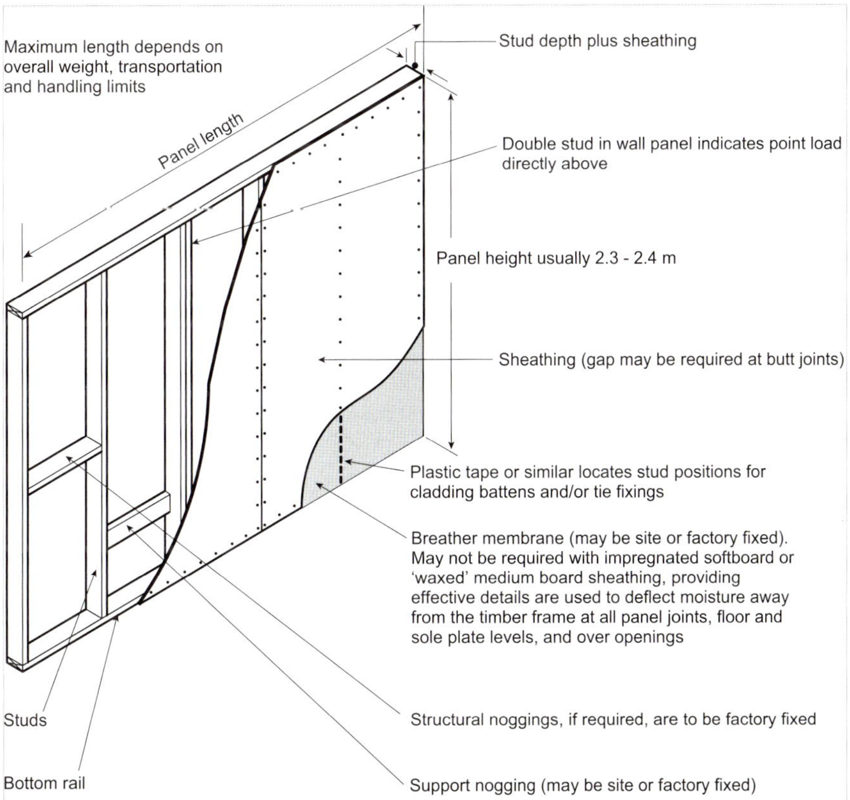

Standard timber frame wall

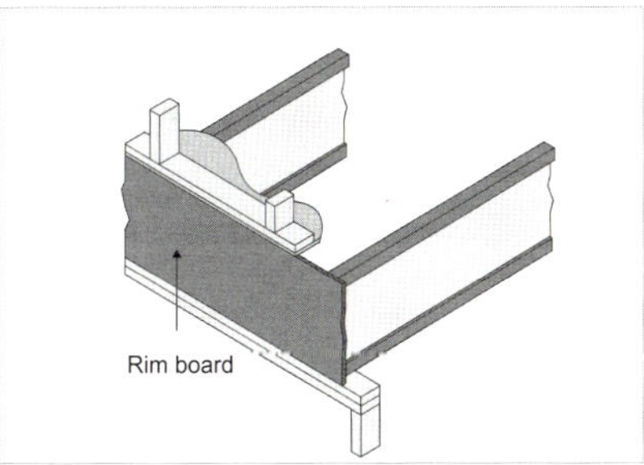

Cassette floor end detail

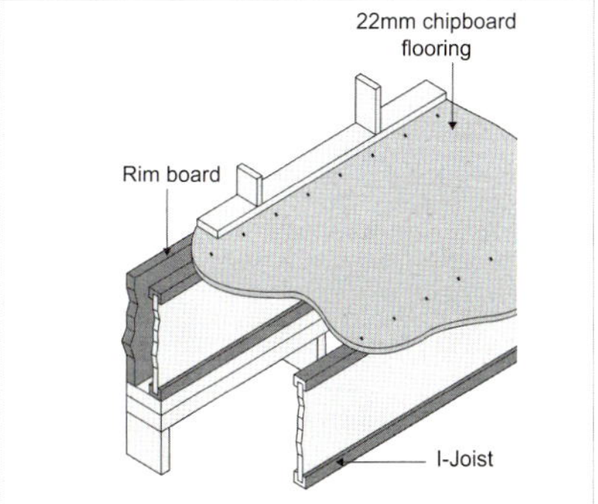

Cassette floor edge detail

Figure 2.16 Standard wall and floor details

Diagrams: Standard timber frame wall, TRADA Technology; Cassette details, Weyerhaeuser

Table 2.2 Illustrative specifications for load-bearing timber frame stud walls for particular performance requirements (IstructE, 2007)

Wall type	Minimum stud sizes (mm)	Plasterboard	Insulation thickness within frame	Fire resistance (minutes)	Thermal trans-mittance U (W/m²K)	Sound insulation Rw (dB)
External	38 X 89	1 layer 12.5mm Type A	90mm	30	0.35	50
External	38 X 140	1 layer 12.5mm Type A	140mm	30	0.31	55
Internal	38 X 63	1 layer 12.5mm Type A	65mm	30	N/A	40
Internal	44 X 75	1 layer 15mm Type D with staggered joints	None	30	N/A	43
Internal, party	38 X 89	1 layer 19mm + 1 layer 12.5mm Type A with staggered joints on each leaf	65mm in each frame	60	N/A	55

NOTES

The values shown relate to walls insulated with Isowool, a glass mineral wool, and are reproduced by kind permission of the manufacturer. Insulating materials and plasterboards are produced in different densities, and the manufacturers should be consulted on the product types to which such test data apply. The insulant is laid directly on top of the plasterboard ceiling unless otherwise stated.

External walls have 1 or 2 layers of plasterboard on the inner face internal walls have 1 or 2 layers on both faces. For thermal and sound values "External walls" assumes 100 mm brickwork or blockwork, a 50mm cavity, and 9mm thick plywood or OSB sheathing on the outer face.

Minimum stud centres 600mm.

Plasterboard types from BS EN 520.

Airborne sound. Improved sound performance can be obtained by the use of proprietary resilient bars.

Type D is a high density, sound-insulating board.

Two separate timber frames spaced 50 mm apart, consisting of timber studs with mid-height noggings. Two layers of plasterboard on the internal face of each frame.

Wall panels and floor systems need to satisfy building performance issues (fire resistance, sound insulation and thermal performance). *Table 2.2 and Table* 2.3 show standard wall and floor specifications for a range of building performance ratings.

The major advantage of timber frame construction relative to the brick and block construction process is speed of construction. The removal of brick laying from the critical path can result in a wind and water tight building envelope in a matter of hours. To realise the full-time saving potential of timber platform frame, a method of construction has evolved that incorporates the off-site construction of the wall and floor components and the on-site preparatory construction of the roofing system (*Figure 2.17*).

Table 2.3 Sample specifications for floor constructions for different performance requirements (IstructE, 2007)

Floor type	Minimum joist depth (mm)	Plasterboard	Insulation thickness	Fire resistance (minutes)	Sound insulation R_w (dB)
Intermediate floor, timber I-joists	240	1 layer 12.5mm Type A	65mm	30	40
Intermediate floor, solid timber joists	195	1 layer 12.5mm Type A	100mm	30	40
Separating floor, timber I-joists	240	1 layer 19mm Type A beneath walking surface. 1 layer of 19mm + 1 layer of 12.5mm Type A beneath the structural floor, staggered joints.	25mm laid on sub-deck between acoustic battens + 100mm laid on plaster-board ceiling.	60	50 airborne 53 impact

NOTES

The illustrative performance values shown are specific to floors insulated with Isowool, a glass mineral wool. They are reproduced by kind permission of the manufacturer. The specifications of plasterboard and insulating materials can change, so always consult the product manufacturers for details of tested configurations relating to the materials currently manufactured.

Gypsum plasterboard types from BS EN 520.

The insulant is laid directly on top of the plasterboard ceiling unless otherwise stated.

Joists at 600mm centres screw fixed to 22mm tongue and groove particleboard.

Joists at 450mm centres screw fixed to 18mm tongue and groove particleboard.

Joists at 600mm centres. The separating floor consists of a floating floor on a structural floor. The floating floor consists of an 18 mm wood particleboard walking surface spot-bonded to 19mm Type A plasterboard supported on 70mm deep acoustic battens at 450mm centres fixed to a 15mm OSB sub-deck. The structural floor consists of the OSB sub-deck nailed or screwed to timber I-joists at 600 mm centres. The ceiling plasterboard is fixed to a resilient bar acoustic channel screw-fixed to the underside of the joists.

Wall panels

Floor cassette

Roof system

Figure 2.17 Crane erection of off-site produced components and on-site preparatory constructed roof system

3 Applying the principles of production to construction

Timber off-site MMC is a combination of off-site production and on-site construction. Depending upon the project (one-off, volume house builder, private or public sector, domestic housing, commercial buildings, flatted development etc) and timber form of construction (timber platform frame, massive timber, beam and post construction etc), a degree of off-site production will take place. This ranges from open panel to volumetric, followed by on-site construction which requires the formed parts to be brought together to create a building envelope.

3.1 The production conversion process

Production as a process is a conversion of an input to an output and the conversion process can be divided into sub-processes, which are also conversion processes (*Figure 3.1*). Minimising the cost of the total process requires the cost of each sub-process to be minimised. The value of the output of a process is associated with the cost (or value) of inputs to that process [19].

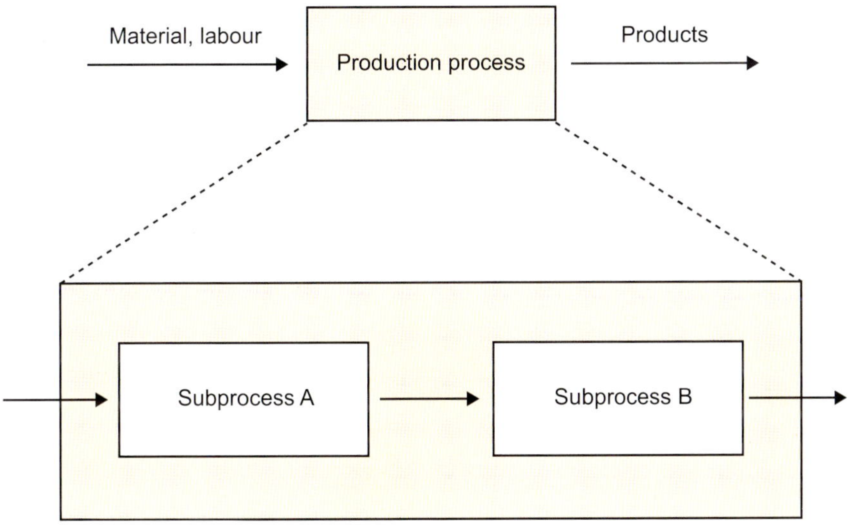

Figure 3.1 The conventional view of a production process as a conversion process that can be divided hierarchically into sub processes [19]
After Koskela

The concept of the production line resulted from this theory and was first exploited in the car manufacturing industry by the Ford Motor Co in 1913, whereby the task was brought to the worker. The production line thus eliminated the necessity to fetch and carry which, in tandem with standardisation of the product, reduced the requirement of the worker to simple, repetitive operations that in turn led to automation of the procedures by machinery.

Taking the construction process to the factory allows the production line process to be applied. The level of construction taken to the factory will depend on the project and market being supplied. For instance if the project or market is large, such as in the case of supplying a volume house builder with a set range of house types constructed from a range of standard component parts, then there is a high degree of repetition. Therefore, it makes sense to take these standardised components such as wall panels to be fabricated in factory environment. It is also important to consider the other components and sub-assemblies as well (ie the formation of a truss rafter or a floor cassette) since production line processes are used when forming these products, although the end product may also be installed individually on-site.

However, if the project is for example a one-off private house builder wanting to use non-standard timber sizes to form a bespoke timber system, then the use of off-site mechanisation on the large scale is limited. Therefore, repeatability and standardisation lend themselves to the off-site environment but this can impinge upon overall flexibility of the system. The ability to achieve a large degree of flexibility in a mass production environment, whilst meeting the requirements of individual customers, is regarded as mass customisation.

Production can therefore be conceptualised as a flow of material and/or information from raw material to the end product. In this flow, the material is processed (converted), is inspected, and is either waiting or it is moving (*Figures 3.2 and 3.3*). These activities are inherently different. Processing represents the conversion aspect of production; inspecting, moving and waiting represent the flow aspect of production. Flow processes can be characterized by time, cost and value. Value refers to the fulfilment of customer or, with respect to construction, client requirements by means of processing and conversion. In most cases, only processing activities are value-adding activities. For material flows, processing activities are alterations of shape or substance, assembly and disassembly [20].

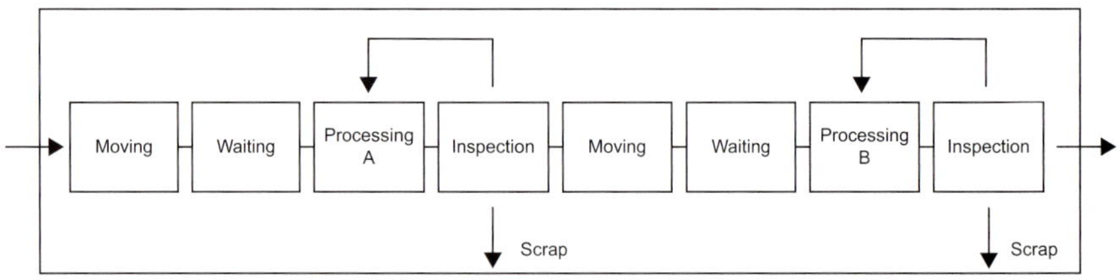

Figure 3.2 Production as a flow process: simplistic illustration. Moving, waiting and inspection represent non-value-adding activities, in contrast to value-adding processing activities. [20]
After Koskela

3.2 Lean production

In *production management* value is created by the flow, both for internal and external customers. Value streams through the process, both within the process itself but also via the supply chains. Bergstrom and Stehn [21] describe lean production as *"a holistic management philosophy, with product quality as the primary goal, which underlines the critical importance of employees, customers, improvements of the two main conversion processes, design and production, and elimination of all other activities, to achieve customisation of high volume products"*. The aim for perfection is the foundation of lean production, a concept that stemmed from the production system of the Japanese car manufacturer Toyota (*Figure 3.4*).

The Japanese word for continuous improvement is "Kaizen". The philosophy of Kaizen places emphasis on every employee at every level in an organisation to make efforts of constant improvement. In terms of lean, these improvements are simple; reduce unnecessary operations (waste) with simple methods to promote increased flow targeted at creating customer value.

Process: cutting of timber studs

Moving: timber studs moved to next station

Waiting: storage of materials prior to process

Process: fabrication of wall panel on bench

Figure 3.3 Timber frame wall production process

In order to achieve value adding flow, the pull concept is endorsed. Waste is to be eliminated to amplify value and so variability and complexity are to be continually reduced. To achieve perfection (the goal of lean production),

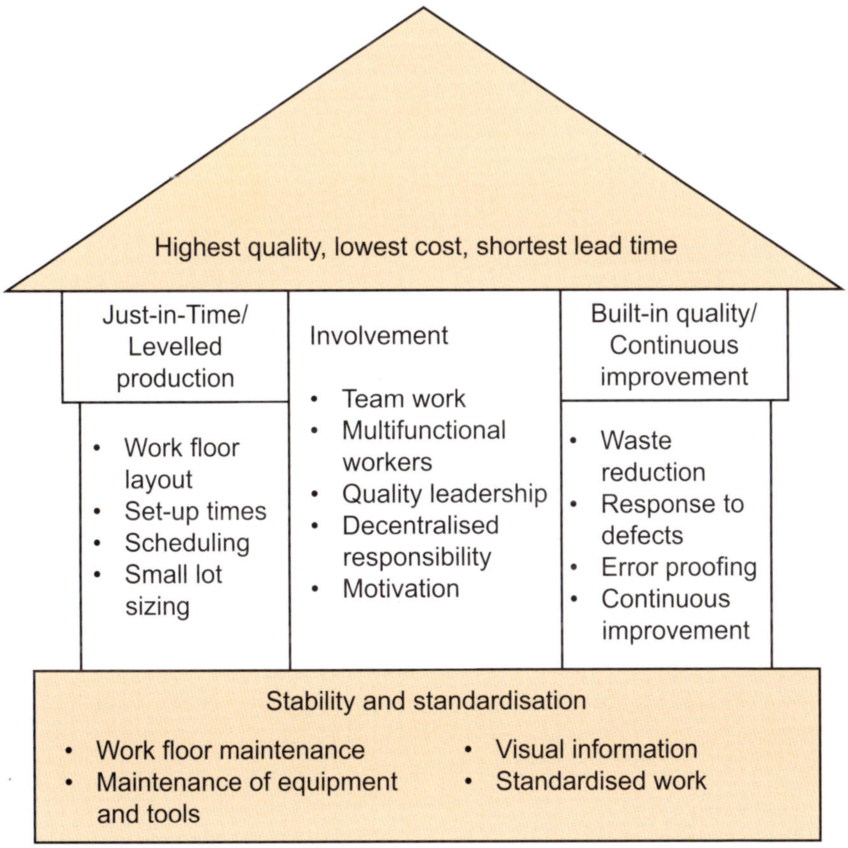

Figure 3.4 The house of lean production [18]
After Hook and Stehn

transparency is required and the involvement of personnel, who are encouraged to identify mistakes. Lean production is therefore a never-ending quest within which constant improvements towards perfection are focused by applying the six basic principles for production improvement. However, one of the most important features of lean production is the success in managing the balance between standardisation and flexibility.

3.3 Six ways to improve the production process

There are key principles that can be applied to reduce waste, lead times, variability and complexity whilst improving flexibility and transparency.

3.3.1 Reduce non-value adding activities (waste)

Waste is not just the unusable material that is a by-product of the production process. Waste falls into a variety of categories such as overproduction, corrections, material movement, processing, inventory, waiting and unnecessary motion. The Japanese have a term, *muda,* that is "an activity that consumes resources but creates no value". Taiichi Ohno, developer of the Toyota Production System, identified seven types of *muda*:

- Overprocessing of parts due to poor processes or product design (*Figure 3.5*)
- Overproduction ahead of demand
- Waiting for the next processing step
- Unnecessary transport of materials
- Inventories (stock of raw material, work in progress and finished parts) that are more than the absolute minimum
- Unnecessary movement by employees
- Production of defective parts and scrap.

Storage of wall panels prior to loading for transport requiring a further process

Direct storage of wall panels on transportation trailer removing additional process

Figure 3.5 Removing sub-processes from the overall process

3.3.2 Reduce the lead time

Lead time is the sum of process time, inspection time, wait time and move time. Reduction or compression of the lead time can be achieved by eliminating non-value-adding time (storage, moving between work stations, downtime etc). For instance, actions such as redoing can be eliminated by ensuring defects do not occur, standardisation will assist with this and will also alleviate process delays. The introduction of the concept of 'pull' (work is released on the basis of systems status or as and when required by the next customer or indeed the end customer or client themselves) rather than 'push' to the production process (release of work is scheduled) (*Figure 3.6*) serves to prevent delays and eliminate wait times. When determining the layout of the production process, "production cells" (production activities which feed each other are placed together) need to be implemented to minimise transport distances.

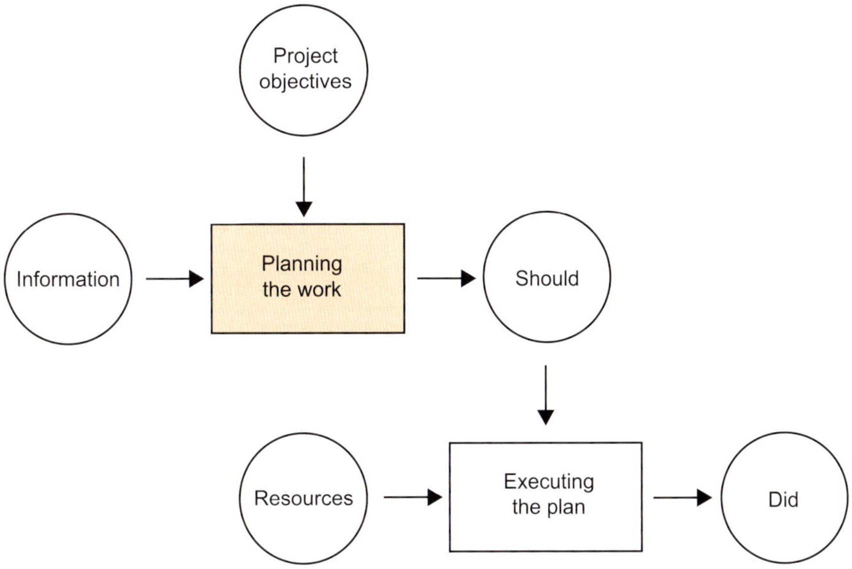

Figure 3.6 A traditional (Push) planning system [22]
After Ballard

The important relationship between work-in-progress (the final product is as yet uncompleted) and lead time (reducing work-in-progress reduces lead time if output remains constant) is demonstrated by the following [23]:

Lead time = work-in-progress / output

By compressing the lead time, customers are satisfied as a result of delivery certainty and speed. Compression also results in improved throughput and this reduces the requirement to forecast future demand, thereby easing stocking requirements and order management.

The following are recommendations to achieve a reduction in lead time:

- Eliminate work-in-progress.
- Reduce batch sizes.
- Change plant layout to minimise movement.
- Ensure things keep moving purposefully.
- Smooth and synchronise flows.
- Reduce variability.
- Change activities from sequential order to parallel order (ie processes that feed the end product are acting concurrently).
- Isolate the main value-adding sequences from the support work.

3.3.3 Reduce variability

There are two types of variability in production flows:

1. Process time variability is the time it takes to process at one work station relative to another as a result of natural variability (minor fluctuations

due to differences in operators, machines and material), random outages, setups, operator availability and rework due to poor quality.

2. Flow variability is the variation in time taken for a job to arrive at a work station.

Variability increases the lead time and decreasing it requires the root causes to be identified and eliminated. Attention therefore needs to be focused on maintenance to minimise outages, to find set-up time reductions, and to establish improvements to product quality that can reduce re-work.

Manual fabrication of floor cassettes and automated wall fabrication of wall panels in off-site timber platform frame construction are two sub-processes which result in the final product (*Figure 3.7*). Both sub-processes have relative variabilities and require good planning within the overall production process to ensure efficiency is achieved.

Floor cassette fabrication

Wall panel fabrication

Figure 3.7 Sub-processes of different speeds feeding the same process

3.3.4 Reduce complexity

Simplification, standardisation (*Figure 3.8*) and minimisation of the number of steps, parts and linkages reduce complexity and, in turn, minimise defects and waste generation. Adding complexity to a product or process only serves

Multi-cripple studs should be secured to each other with 3.1mm dia.x 64mm long galvanised ringshank nails at 400mm centres staggered mid distance between edge and centreline with no nail closer than 60mm to end of studs.

Lintels should be secured to each other with 3.1mm dia. x 75mm long galvanised screws at 300mm centres staggered mid distance between edge and centreline with no screw closer than 60mm to end of lintel

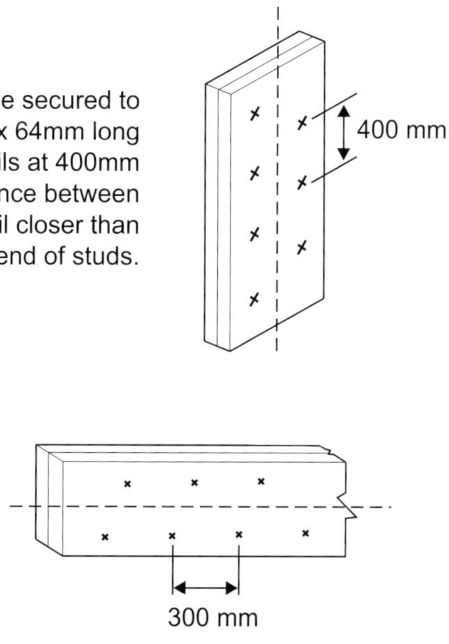

Figure 3.8 Standardised nailing schedule of timber components

to increase the final cost. Therefore, in terms of overall value to the customer, is the additional complexity worth it or can the process be simplified? Complexity also impinges upon reliability: complex systems are inherently less reliable than a simple system.

Simplifying the process can be achieved by reducing the number of components required or by eliminating non-value-adding activities from the production process. Simplification should not be restricted to the production line; organisational structure has an influence on out-put and, where possible, should also be simplified.

Practical approaches to simplification include: shortening the flow by consolidating activities; reducing the part count of products through design changes or prefabrication of parts; standardising parts, materials, tools etc; decoupling linkages; and minimising the amount of control information needed.

3.3.5 Increase flexibility

Historically, off-site construction had a poor public perception and a primary reason for this was lack of consumer focus which resulted in a production focussed product with no individuality. Flexibility is required in order to achieve individuality or customisation. There are four basic groups of manufacturing flexibility:

- mix flexibility (number of different products produced)
- new product flexibility (speed of product induction)
- volume flexibility (ability to vary production)
- delivery time flexibility.

In order to achieve flexibility without impinging upon simplicity, practical approaches can be applied such as minimising lot sizes to closely match demand, reducing the difficulty of setups and changeovers, customising as late in the process as possible, training the multi-skilled workforce, training the workforce in operational flexibility, and using general purpose machines.

In off-site MMC, flexibility is important in order to achieve architectural requirements such as for example varying house types and layouts (*Figure 3.9*).

Figure 3.9 Standard house types for off-site construction demonstrating flexibility
Drawing: Oregon

Early in the design stage it is necessary to consider how best to rationalise the component parts being used in order that they can still provide the required level of variation, taking account of the customer or clients requirements. For example different forms of construction (panel system, post and beam, massive timber) will have a different available range of variations based on the relative level of component parts and what they can accommodate in terms of section size, strength and connectivity. From a structural perspective it is important early on in the procurement process to consider for example the different load span conditions of the system as this will dictate the type and depth of section required and the relative level of standardisation given the security of supply and relative cost.

3.3.6 Increase transparency

The primary focus of making the production process and associated systems transparent is to engage the workforce in a manner that will motivate improvements and increase control through responsibility. Transparency makes the flow of operations from start to finish visible and comprehensible to all employees. This can be achieved by making the process directly observable through organisational and physical means, measurements and public displays of information.

Practical approaches to enhancing transparency include:

- establishing the basic housekeeping rules to eliminate clutter
- making the process directly observable through appropriate layout and signage
- standardisation
- rendering invisible attributes of the process visible through measurement
- embodying process information in work areas, tools, containers, material and information systems
- utilising visual controls to enable any person to immediately recognise standards and deviations from them
- reducing the interdependence of production units (focused factories).

Figure 3.10 Increasing transparency

Figure 3.10 shows tools stored on a board with a standardised template designating the storage position of each piece of kit. This ensures a clutter free environment and transparency. Workers know instantly what tools are available during the execution of a work task or what needs to be reinstated at the end of a work task.

3.4 Modern Information and Communication Technology (ICT)

Industrialised processes require accurate and reliable information. Modern ICT provides tools that effectively handle updates and changes of digital material and provide solutions for information exchange and data storage. An extensive use of modern IT-tools supports the different processes by enabling more accurate documents and hence good conditions for an effective production in which errors are discovered early and problems in the manufacturing and assembly phases are avoided.

The design process can be linked to the manufacturing process using computer-aided design/computer-aided manufacturing (*CAD/CAM*).

These computer systems are used to design and manufacture products. Architectural layouts produced in 3D CAD environments can, for example, automatically create cutting lists for the the production process and this can be automated using an optimisation saw, thereby reducing wasted. In more high-tech applications the CAD/CAM system can be used for computer numerical control (CNC) machining such as for the routing out of service channels or the formation of proprietary metal work connectors. There are now *4D CAD* systems, capable of linking components in *3D CAD* models with activities from the design, procurement, and construction schedules.

The structural engineering design of standard timber truss and floor systems are normally carried out using the computer software of the manufacturer and supplier again automating the process. *Whole house engineering software* capable of carrying out full combined system (roof, walls and floors) analysis and design is also becoming more prevalent and this will further optimise the structural engineering process. *Finite element analysis* (FEA) software can be used where further structural optimisation is required. FEA carries out detailed structural analysis and this can be particularly useful for connection design.

From a building performance perspective in the UK, the *Standard Assessment Procedure* (SAP) is used to assess the energy performance of buildings based on the energy costs associated with space heating, water heating, ventilation and lighting, less cost savings from energy generation technologies. SAP software is used to automate this process and provide an energy ratings for the dwelling based on an extensive range of inputs.

From an overall planning perspective there is the opportunity to use *Enterprise Resource Planning* (ERP – a comprehensive planning mechanism that is supported by information technology based systems to manage parts of or the whole supply chain [(10) (24) (25)]). ERP adoption is essential for the success of off-site MMC.

ERP is regarded as one of the most innovative developments in ICT of the 1990s [(25)] as it implements best practice procedures, standardises business process and systems and provides an ideal platform for transparency. ERP was traditionally used by capital-intensive industries, such as manufacturing, construction, aerospace and defence, although more recently ERP has been used in the finance, education, insurance, retail and telecommunication sectors.

The requirements of *Total Quality Management* (TQM) and *Business Process Engineering* (BPR) are considered to be the key drivers behind the use of increased levels of ICT for key business functions. However, although the adoption of ERP implements some of the concepts of lean it has not always been successful and one of the major challenges is assuring future flexibility, ie the extent to which the system can be reconfigured to accommodate new business models or processes. To achieve the right balance and therefore the successful use of ERP, the implementation process is critical especially in terms of buy-in since one of the main obstacles facing ERP – particularly within the construction sector - is resistance to change.

To achieve successful implementation the following list of guidelines has been developed [26]:

- Infrastructure resource planning – make sure that adequate infrastructure is planned for in such a way that it becomes reliably available in good time (both for the pre-implementation and the post-implementation stages).
- Local area network – ensure network support for any ERP or other applications.
- Servers – deploy adequate server/network, even during the training modelling phase.
- PCs – introduce new PCs with the latest configuration sufficient for most ERPs.
- Training facilities – establish an adequate training centre to work as a competency centre.
- Human resource planning – focus on building a teamwork environment, where team size spans across the entire organisation.
- Education about ERP – carry out ERP education across the organisation to highlight ERP success and failure practice.
- Commitment to release the right people – ERP is recognised as a difficult but necessary project, and the best people must work full-time on the project.
- Top management's commitment – top management must have a change mindset through learning at all levels.
- Commitment to implement – ensure minimal customisation and quick implementation.
- Reasonably well working manual systems – carry out audit exercises to find the current status and their corresponding correction actions.
- Strategic decision on centralised v decentralised implementation.

When correctly implemented, ERP has been shown to be successful although a high degree of investment is required. Research [21] has demonstrated that ERP can fulfil the needs of off-site industrialised housing if the following pre-requisites are matched:

- There must be external and internal driving forces for the necessary change process to start, with a balance struck between the required level of customisation and efficiency in production.
- The change process serves two functions. First, both external and internal requirements must be understood in order that any implemented system will satisfy the intended needs. Second, the change process must be managed in a way that facilitates ERP implementation.
- The change process must aim for a production system and business processes that are functional from two aspects: meeting the external and internal needs and facilitating ERP system implementation.
- From an organisational perspective the work force require to understand ERP and how it can contribute to meeting both external and internal needs, ie simply implementing a software system to improve efficiency will not work.
- During the ERP implementation process, and when ERP is in use, it is critical that the day-to-day work is not hindered. For an SME, careful pre-implementation will facilitate a flexible installation process and smooth transition period, ie ERP implementation is secondary to day-to-day work as well as the initial change process implementation.

It is often the case in timber construction that the overall process is fragmented with different parts of the project outsourced including strucutral design, building performance analysis, component manufacture and construction. A fragmented approach can impinge upon the communication of critical information resulting in substandard design and subsequent low quality product. The use and integration of sophisticated sofware tools can, if correctly implemented and endorsed, resolve this.

An integrated platform for design, manufacture and construction connecting the lifecycle of a project – conceptual planning, architectural design and detailing, structural design and detailing, manufacture, construction and project management – would make the whole process transparent and capable of being monitored. As a result design can be optimised, self-improving and adjustable to a client's project specific requirements and Enterprise Resourced in real-time. Information could be securely created, shared, and synchronized with the extended project team. A holistic modern system would close the information loop and therefore result in lean design procurement, improve the decision making process and remove unnecessary steps from the critical path to full design completion.

3.5 The construction conversion process

Whether traditional or modern methods are being applied, construction is fundamentally a set of activities aimed at generating a certain output and construction can consequently be considered to be based on the conversion model. However, on a traditional construction site there is a lack of flowing production that can be measured and improved. *Construction management* is the organisation and running of a construction project, as a result the construction manager is often faced with the difficulty of trying to manage the work of temporary organisations on what are regarded as a transformation activity. Transformation can be regarded as taking place only once (although in practice there is a degree of repetition when considering the transfer of applied knowledge as a result of carrying out similar processes on other sites) and as a result the process is difficult to gradually improve (*Figure 3.11*). Further to this as a result of the often fragmented nature of the industry, especially in the UK due to the reliance on sub-contractors, the incentive to improve the overall process is diluted. Therefore, it is difficult to create added value during the construction conversion process.

Figure 3.11 A typical 'chaotic' construction site

3.5.1 Lean construction

In many respects construction management is different to production management; with construction being a project-orientated framework it lacks the long-term striving for perfection that is the goal of lean. Construction, being project-focussed, can be regarded as a temporary production system that is normally linked to other temporary and permanent production systems for materials, equipment and labour. Lean production principles consequently do not transfer directly to on-site construction, but what has been proposed is the transformation, flow, value-theory or TFV-theory [19]. The introduction of transformation as an element in lean theory reflects the construction industry's idea of an object being gradually enhanced by craftsmen whose work is not necessarily organised in a flowing manner.

A range of techniques can be applied to achieve lean construction. These include the implementation of the Last Planner System, improving project visualisation, engaging more efficiently with the work force through daily huddle meetings or tool box talks and improving site efficiency by means of improved "housekeeping" methods.

3.5.2 Last Planner System

The Last Planner System™ (LPS™) is a lean-production-based project planning and management system, conceived by the founding members of the Lean Construction Institute (www.leanconstruction.org), which recognises the interdependency of the associated trades of a construction project with a view to implementing:

- robust communication methods
- multi-trade collaboration to enhance project predictability
- work flow through pull implementation.

LPS™ has four key elements [(27)]:

- Collaborative programming – creating and agreeing the production sequence (and compressing it if required) collaboratively.
- MakeReady – making tasks ready so that they can be done when they require to be done.
- Production planning – collaboratively agreeing production tasks for the next day or week.
- Continual improvement – learning and improving the project-planning production process.

The LPS™ works in addition to the traditional general project planning (master plan). Intermediate and weekly task plans are introduced and project slippage is followed up by applying a percentage plan complete (PPC) indicator.

The system can be understood as a mechanism for transforming what SHOULD be done into what CAN be done.

SHOULD, CAN and WILL are defined [(22)];

SHOULD – work that is required to be done according to scheduling requirements.
CAN – work that can be accomplished on account of various constraints on the field.
WILL – work commitment which will be made after all constraints are taken into account.

CAN be done tasks are considered once the constrictions (bottle necks) have been analysed and eliminated and are called "assignments." The person or working team that outlines them is called "Last Planner", from which the system derives its name (*Figure 3.13*). The last planners form a team to develop a schedule that works backwards from the completion date known as Reverse Phase Scheduling (RPS) by introducing pull into the systems. Assignments from the RPS are then scheduled to form a Weekly Work Plan of tasks that WILL be carried out. The production unit function is to carry out the assignments correctly by means of a continuous learning process and corrective action. The PPC indicator measures the planning and the productivity performance of the production unit. It is represented as a ratio of completed assignments divided by planned assignments. A good performance is above 80% and a poor one is below 60% (28).

Just as it is necessary to determine the percentage of completed activities, it is also important to establish the reasons why the working assignments were not fulfilled. This provides the information required for the improvement of PPC, thus facilitating a more efficient completion of the project.

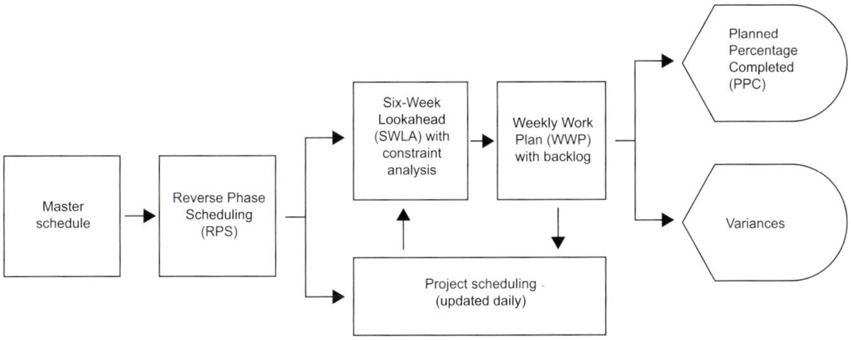

Figure 3.12 The sequence of the Last Planner Process (29)
After Salem

Figure 3.13 shows how Last Planner changes the traditional planning system.

Application of the LPS™ to projects has shown that simultaneous improvement in all four dimensions cost, schedule, quality, and safety is possible:

- Reduced project cost: workers spend less time waiting for work or working around problems.
- Reduced project duration: available work does not sit idle waiting for workers.
- Improved quality: work is done in its natural sequence and is released from one participant to the next only when it meets established criteria.
- Improved safety: the work environment is more stable so fewer ad-hoc efforts are required.

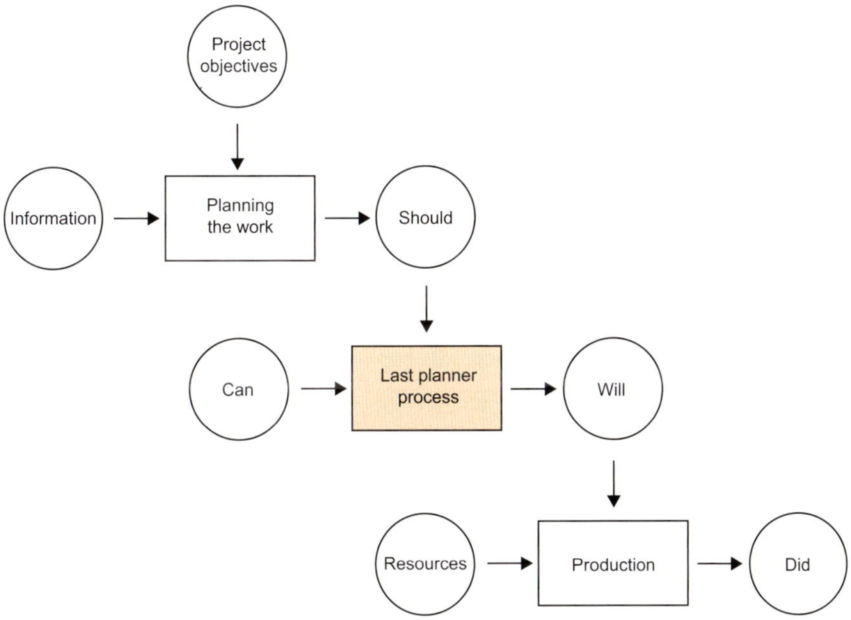

Figure 3.13 Last Planner System™ within the production framework (22)
After Ballard

Software tools have been developed in order to guide production units, in accordance with the Last Planner methodology, to create a reliable work plan. The developed software includes *WorkPlan*, which is for specialist contractors to develop weekly work plans. WorkPlan is a standalone program but it also allows for production units working on construction projects to import basic schedule information, such as a list of activities and precedence relationships, from *Microsoft Project*. Further to this there is *DePlan* software that combines WorkPlan with *ADePT* (30) and has the ability to represent design process models, perform dependency structure matrix analysis, and develop design programs for the overall project and individual disciplines. DePlan guides production units working on design projects to import schedule information from ADePT.

WorkPlan has been extended into *WorkMovePlan* (*Figure 3.14*), which includes capabilities for distributed planning and coordination and space scheduling (31). The distributed planning and coordination capabilities allow production units to increase the reliability of their plans by sharing work package, space scheduling information, and constraint information. The space scheduling capabilities allow each planner to explicitly allocate space, including workspaces, laydown areas, storage areas, and access paths.

The LPS™ takes fear of losing control away from the top management team as a result of improving transparency, a key lean philosophy. Further to this LPS™ can result in the project team being more confident in taking

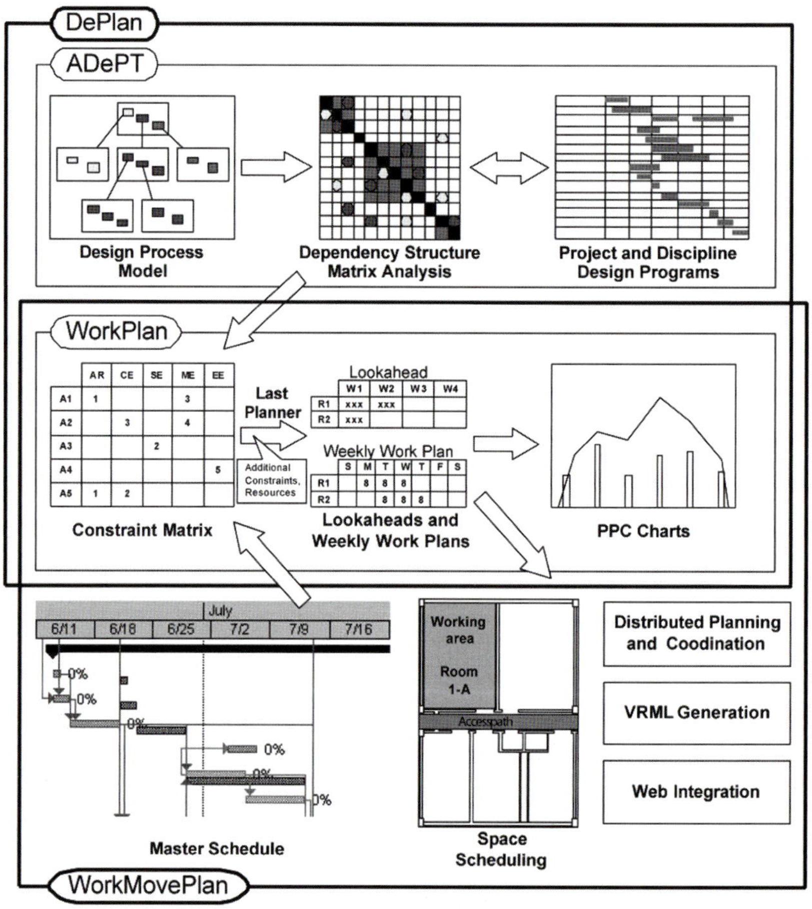

Figure 3.14 Last Planner Software including WorkPlan and WorkMovePlan (31)
After Choo and Tommelein

project responsibility and can also make improvements to the work culture of the organisation.

3.5.3 Increased visualisation

To increase the level of transparency on-site to engage the work force, key information can be communicated around the construction site through posting signs and labels. In lean manufacturing this tool is known as "Visual Controls" and is used to remind workers of project elements such as workflow, performance targets and specific required actions. This includes signs related to safety, schedule, and quality (*Figure 3.15*).

Figure 3.15 Examples of increased on-site visualisation from Japan

3.5.4 Daily huddle meetings (tool-box meetings)

Engaging the work force is a key concept of lean and two-way communication is the key to this. In order to effectively achieve this on a construction site, daily huddle meetings can be used. These consist of a brief daily start-up meeting where team members quickly give the status of what they have been working on since the previous day's meeting, especially if an issue that has been given might prevent the completion of an assignment. Further to this the meetings can be used for work force training and project progress reports (*Figure 3.16*). This tool is similar to the lean manufacturing

Figure 3.16 Tool box talk demonstrating the correct installation of timber concrete shear connectors and associated strength

concept of employee involvement that ensures rapid response to problems through empowerment of workers and continuous open communication through the tool box meetings. One of the key reasons for the meetings is to maintain employee satisfaction (job meaningfulness, self-esteem, sense of growth).

3.5.5 The 5S Process (Visual Work Place)

The 5S Process, also referred to as the Visual Work Place, is a tool that first saw inception in Japan and which is used in lean manufacturing to ensure good housekeeping on five levels:

- Seiri (sort) refers to separate needed tools/parts required and the removal of unneeded materials (trash).
- Seiton (straighten or set in order) is to neatly arrange tools and materials for ease of use (stacks/bundles).
- Seiso (shine) means to clean up.
- Seiketsu (standardize) is to maintain the first 3Ss by developing a standard 5Ss work process with an expectation of system improvement.
- Shitsuke (sustain) refers to the creation of the habit of conforming to the rules.

Construction is more dynamic than manufacturing and a continuous effort that includes all trades is therefore required to implement the 5S Process. However, the culture of construction can be an obstacle to the implementation of the 5S Process, with construction workers normally often taking the view that they are employed to conduct a physical process and are thus not committed to housekeeping. The 5S process can therefore be the most difficult to implement; however, successful implementation can result in improved safety, productivity, quality, and set-up-times improvement, creation of space, reduced lead times, cycle times and increased machine uptime.

3.6 Innovation

Research has revealed that organisations that evolve a framework for long-term innovation react faster to threats or surprises, are more likely to take effective action and are better at sustaining the momentum of change (*Figure 3.17*). As such, they are more likely to succeed [32].

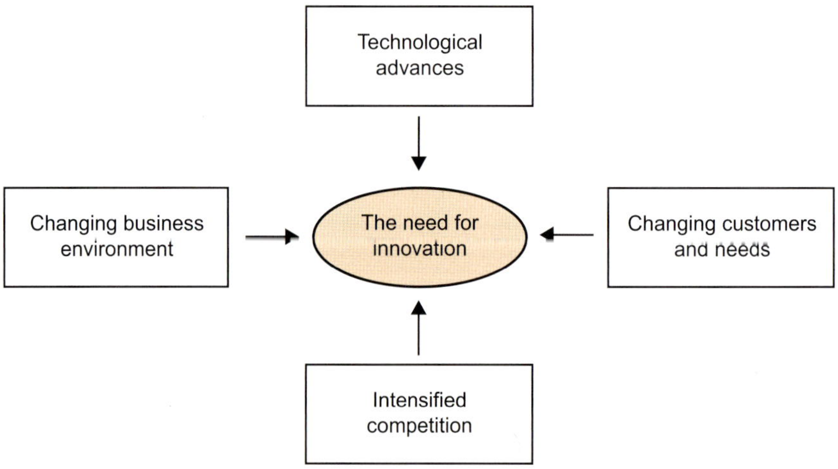

Figure 3.17 Drivers of the need for innovation [33]
After Goffin and Mitchell

Innovation is about the successful exploitation of new ideas to improve processes, products, techniques or use of materials. In terms of process it is the introduction of changes that are significant departures from the usual way of doing things. From a product perspective it is the transformation of an idea into a novel product, operational process or new service. When considering techniques or new materials, it is the employment of design or construction techniques or materials that do not have a proven history of performance or are not covered by an organisation's current practice. There is therefore a degree of risk associated with innovation. However, without innovation, companies will quickly lose their competitive edge.

For consumers, innovation means higher quality and better value goods, more efficient services (both private and public) and higher standards of living. For businesses, innovation means sustained or improved growth. The innovative company or organisation delivers higher profits for its owners and investors. For employees, innovation means new and more interesting work, better skills and higher wages. Equally, an absence of innovation can lead to business stagnation and a loss of jobs.

Off-site MMC is a lean approach to delivering built systems. Timber has evolved as a construction material. However, this evolutionary process has to continue and, to do so, there is a need for innovation. Innovation is required from an environmental perspective with the challenge being to become a low carbon economy with reduced waste. Essentially there is a requirement to find new ways to break the link between economic growth and resource depletion and environmental degradation [(34)], and construction has a key role here, given its impact on economic growth and use of natural resources.

Innovation is also required when considering the drive towards more sustainable building envelopes with the need to improve overall energy efficiency, without impinging upon the practicalities of the manufacturing and build process or overall structural robustness. Methods for delivering more energy efficient buildings include the use of new products to improve thermal efficiency, detailing solutions to reduce thermal bridging (heat flow across a poor insulator) and improve airtightness, and the provision of energy from renewable sources. Applied research and development is needed to deliver on these requirements. The application of these newly derived products and process carries with it a degree of associated risk but, in order to adapt and change to meet the challenges of a low carbon economy, it is necessary.

With regard to airtightness and thermal bridging, there are methods available to test the built solution and therefore monitor performance once constructed. For example, thermography is the use of infra-red technology which can be used to create thermal images of built systems and therefore identify areas or particular details where thermal bridging is taking place. These could be repeating thermal bridges such as wall ties or non-repeating thermal bridges such as lintels. Once identified, new and novel solutions can be developed to improve the thermal performance.

The implementation of novel solutions, such as reducing cold bridging, to improve thermal performance can have an effect on structure. Wall ties, for example, tie the masonry skin to the timber system and lintels support loads over openings. Therefore, structural analysis is required in order to ensure thermal performance improvements do not impinge upon overall structural performance. The structural analysis may be a simple design check or it may require further more complicated analysis using sophisticated design software such as finite element analysis modelling or indeed laboratory testing.

The inclusion of new and novel solutions must be balanced with the associated impact on the manufacturing and construction processes.

The adoption of new and novel solutions to improve performance requires a holistic approach in order to optimise cost, building (thermal, acoustic etc) and structural performance as well as manufacturing and construction efficiency. Sustained investment is required in order to achieve a multi-disciplinary approach to research and development.

3.7 Case Study: Timber platform frame crane erect method

Oregon Timber Frame Ltd is a Scottish company specialising in the design, manufacture and erection of structural platform timber frames for the residential market. Oregon is a prime example of a company that endorsed and, as a result of research and development, applied the concepts of the 1998 Construction Task Force report [3] as well as understanding and implementing the concepts of MMC.

In order to continually improve its approach to production and construction Oregon has worked in-house and collaborated with external partners to innovate. This approach was showcased by their Department of Trade and Industry (now the Department for Business, Innovation and Skills) award winning work with Edinburgh Napier University in 2006, which resulted in product improvement through applied research, increased profitability and follow on research activates.

The crane-erect method (*Figure 3.18*) of construction is showcased in this case study as it highlights the success of collaborative applied research in improving the efficiency of the construction process. It also demonstrates

Roof constructed on ground floor slab

Roof lifted out of position

Ground floor panels erected

Cassette flooring installed

1st floor panels installed

Roof system craned into position

Figure 3.18 Crane erect method of timber frame construction

how the project management elements are understood, how information is made transparent to the work force and, importantly, how the research work is a continuous process. These principles are in-line with the concepts of lean construction and the applied research conducted has led to improved quality and efficiency.

The development of a Best Practice Procedure (BPP) had three key drivers:

- health and safety
- speed of erection
- cost.

Of the three key drivers, health and safety was deemed as being the most critical on the basis that effective planning for health and safety is essential if projects are to be delivered on time, without cost overrun, and without experiencing accidents or damaging the health and safety of site personnel [35].

Although preparatory construction of the roof system at ground level reduces the time spent working at height – one of the single biggest causes of casualties in the construction industry – there is the risk of the roof system failing during lifting operations. To reduce the associated risk of failure whilst being lifted into position, an analytical model was developed to analyse the behaviour of a roof system under lifting conditions. However, due to the nature of the support conditions when analysing a roof system under lifting conditions (one point of support would result in a mechanism forming), extensive laboratory testing was required to verify the developed model (*Figure 3.19*) [36]. Sheathing was not applied in order to test the worst case lifting scenario. Sheathing would result in diaphragm action providing stability to the roof trusses under lifting conditions.

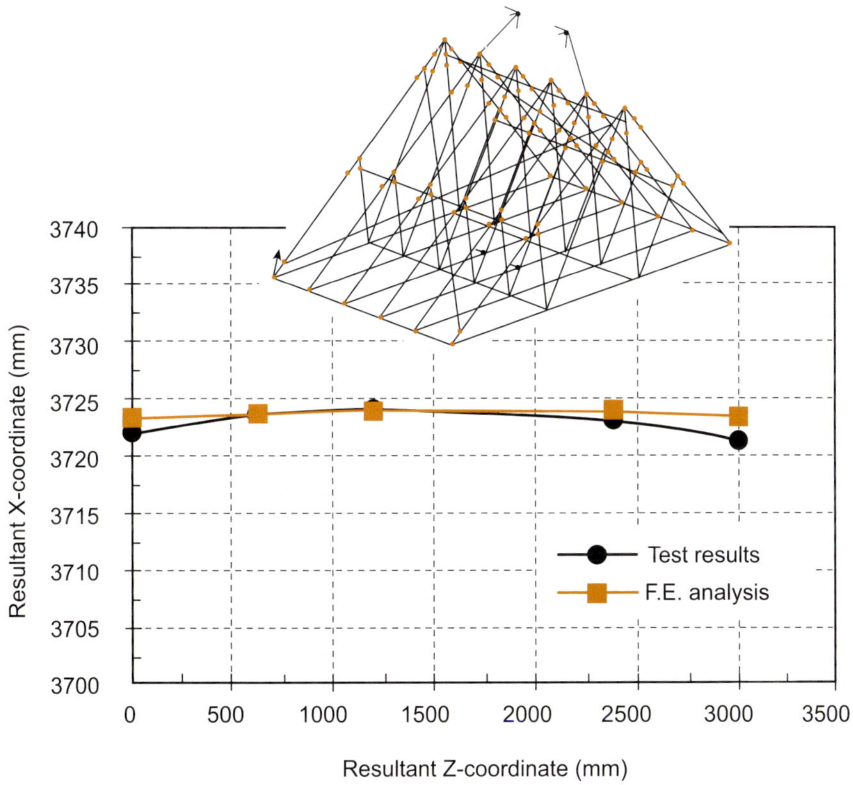

Figure 3.19 Correlation of test node displacement results with analysis

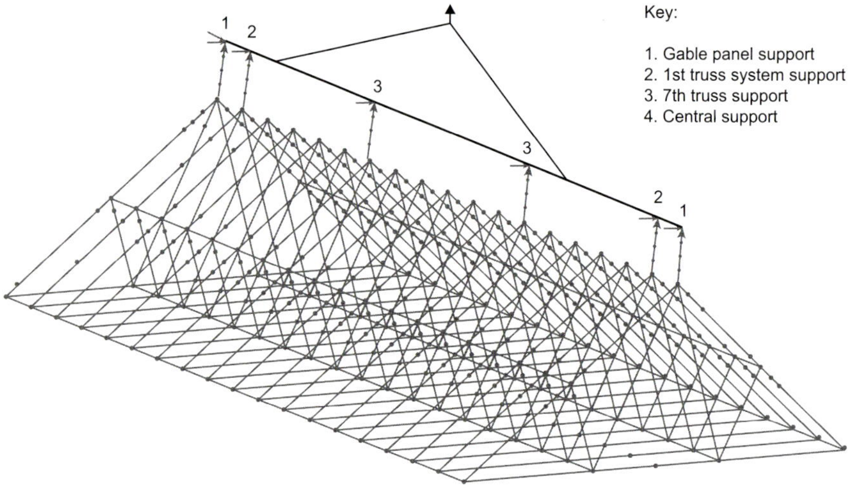

Figure 3.20 Designated apex lifting points for large scale systems using a spreader bar

Using the verified model, a range of methods for lifting roof systems were analysed and two of those derived can be used depending on the nature of the system:

1. Apex point lifting from a spreader bar for large scale or complicated systems (*Figure 3.20*).
2. Upgraded longitudinal bracing lifting for standard cases (*Figure 3.21*).

Figure 3.21 Upgraded longitudinal bracing lifting of standard roof

Standardised information and procedures were produced to ensure that the information from the research work conducted was documented, disseminated and also robustly communicated to contract management teams and erectors. The documentation included guidance notes such that appropriate lifting points could be allocated on drawings at the design stage to ensure correct allocation on-site. In conjunction with the information contained on the drawings method statements were also created and used in on-site "tool box talks" such that the process was fully understood by those involved.

Oregon's research work to improving the construction process is ongoing. In terms of efficiency, savings can be made by considering the sequence of on-site operations relative to the loading sequence of the components on the lorry. The main things that impinge on this are the stability and safety of the load for transportation. On site, safety procedures can be improved and the construction critical path reduced by innovatively improving edge protection methods by eliminating scaffolding requirements. To achieve this, temporary edge protection attached to the floor cassettes has being developed as a further advantage of the crane-erect method (*Figure 3.22*).

Figure 3.22 Temporary edge protection system

3.8 Case study: Mass customisation

Open building manufacturing (*Figure 3.23*) is a new paradigm for building production and procurement. The theory behind open building manufacturing is to combine highly efficient manufacturing techniques in factories and on construction sites with an open system for products and components to offer diversity of supply and building component configuration opportunities to the open market.

The recent ManuBuild project was founded to initiate a shift in construction from being mainly "craft and resource based" towards a future state

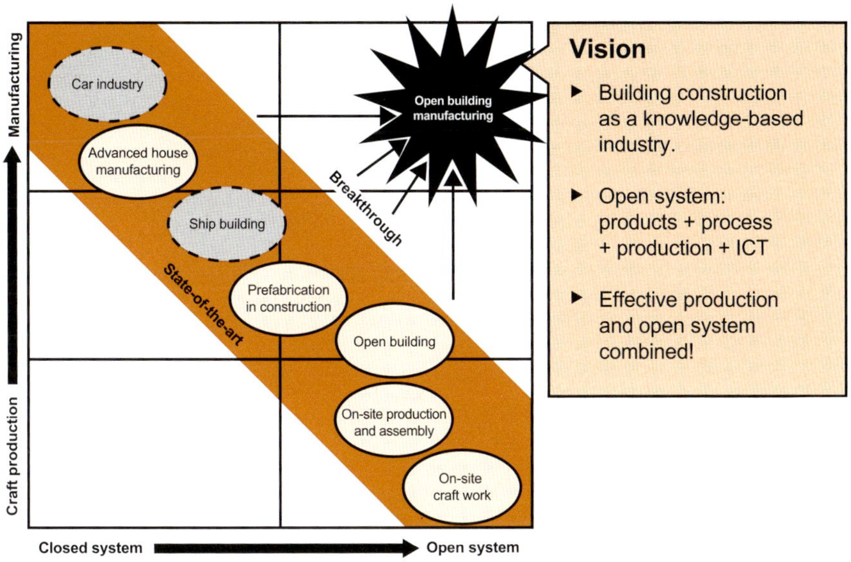

Figure 3.23 Open building manufacturing (37)
After Kazi and Eichert

where it can be characterised as "agile, value-driven and knowledge based". ManuBuild was a 4-year project which commenced in April 2005. Its purpose was to develop novel building concepts, business processes, production technologies and ICT support that can be integrated into one holistic approach, forming the ManuBuild Open Building Manufacturing System.

The success of the Ford Car Company was based on standardisation where only one variant was produced. The car industry has moved on extensively since then. Because of the use of platforms, customers today can select the specification of their own cars. This capability to deliver customer-tailored cars increases the customer's perceived value of the car, while the company can still leverage the economy of scale of mass production. In this way, car manufactures have addressed what is regarded as the "long tail" by applying mass customization strategies (*Figure 3.24*) (38).

In many respects construction similarly tailors unique buildings to its customers, although in the volume house-build market there is a degree of repeatability with the production of house type ranges. These are constantly evolving and changing depending on site specifics, customer or "kerb-side" appeal, regulatory requirements and architectural or client needs (social, private or self-builder). Therefore, construction predominately follows the "individual customisation" strategy. Compared to this, the aim of the ManuBuild initiative was to move to a higher degree of standardisation without losing the ability to deliver customised products.

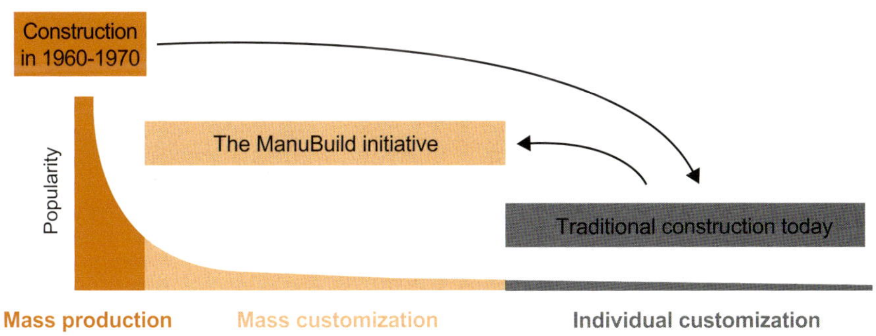

Figure 3.24 Long tail principle (38)
After Thuesen and Jonsson

In this way the ManuBuild initiative was working within the mass customisation paradigm and the deliverables of the project were to assist in the endorsement of this concept (*Table 3.1*).

Table 3.1 ManuBuild Open Building Manufacturing Outcomes (www.manubuild.com)

Methods for the design and specification	Products	Business models and processes	Manufacturing methods and techniques	IT system architecture and tools	Training solutions	Demonstrator Projects
• Methodology for product development • ManuBuild system framework • Architectural typology	• New multi-function materials • Connections and interfaces • Smart components • Partition wall systems • Building acoustics classification • Multi-function and integration • ManuBuild Building Systems	• ManuBuild business model • Reference business process • Organisational models • Service models • Business models • Performance metrics • Market analysis methodology	• Manufacturing methodologies and principles • Process technologies • High volume fixed factory manufacturing • Mobile factory concept • Logistic solutions • Rapid connection and assembly methods	• ManuBuild system architecture • Open ManuBuild system platform • Product modelling ontology • Intelligent catalogues and building templates • Catalogue server • Design configurator • Sales configurator • Logistics and assembly planning	• Training courses • Teaching factory • Viritual reality interactive training environment for off-site manufacturing • University education on off-site manufacturing	• Low rise housing in the UK • Medium rise apartments in Madrid • Medium rise apartments in Sweden

ManuBuild was a European-wide industry-led research project which considered all construction materials and project types. By contrast, an individual example of a timber system that uses standardised components to produce customised buildings is that of Lukas Lang Building Technology based in Vienna, Austria. The component construction method of Lukas Lang Building Technology is based on the idea of 100% industrial production of all building components for up to three-storey ready-to-occupy buildings – from the supporting structure to the building services.

Depending on the creativity of the planner, small construction elements can be individually used for a wide range of buildings with up to three floors. As in the car industry, the production of all parts is outsourced and all assembly work on the building site can be carried out without specialist skills (*Figures 3.25 and 3.26*).

Each building component remains unchanged by assembly work, enabling the building to be modified at any time while the elements retain their value. Owing to their flexibility the components fulfil their function over many generations. Sustainability is thereby ensured for work and material through

Conceptual design — Build process with standardised components

Figure 3.25 Lukas Lang, from building concept to construction
Photos: Lucas Lang

Figure 3.26 A finished Lucas Lang building
Photo: Lucas Lang

longevity. With change of use (family growth or new occupant requirements) the building can be adapted rather than demolished.

A computer-aided design tool acts as a platform for the end user to produce a customised conceptual design based on the standard components that reflects their requirements. This form of construction is therefore *Open Building Manufacturing* and reflects the concepts of the ManuBuild project and what can be achieved using timber as an off-site MMC.

4 Environmental performance and sustainability

4.1 Sustainable development

The planet is increasingly showing evidence of stress caused by environmental damage and resource depletion. How this affects the UK was exemplified in the Energy White Paper [39] which reported that the UK will, at current rates of consumption, be dependent on imported energy for three quarters of its total primary energy needs by 2020 due to depletion of indigenous resources.

Current energy use in the UK is non-sustainable and, as a consequence, there is a drive to make a 60% reduction in carbon dioxide generation by 2050, as part of policy to develop a more sustainable society. Sustainable development is not limited to energy consumption: it is also about ensuring a better quality of life for everyone, now and for generations to come. It includes three broad components: social, environmental and economic often referred to as the 'triple bottom line'.

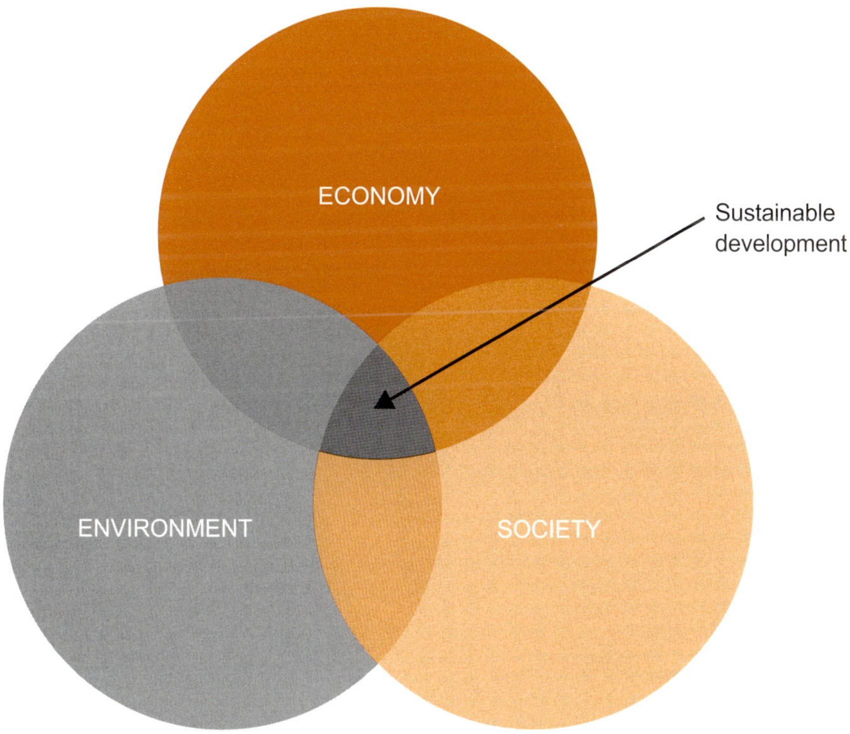

Figure 4.1 Themes of sustainable development
After Shelbourne [40]

Sustainable development consequently requires multidisciplinary, international and interdisciplinary collaborations using both technical and humanistic approaches. It requires a language that is comprehensible by all people, and an ethical stance that often requires the confidence to depart from the norm. Fundamentally, sustainable development is about meeting the needs of the present without compromising the ability of future generations to meet their own needs. In order to create sustainable communities, design – and particularly engineering design – needs to be at the centre of a holistic process that uses all the technical and creative skills of human ingenuity.

The creation of a sustainable society needs therefore to address key issues:

- Social progress that recognises the needs of everyone
- Effective protection of the environment
- Prudent use of natural resources
- Maintenance of high and stable levels of economic growth and employment.

4.2 Sustainable construction

Construction has a significant effect on quality of life: its outputs alter the nature, function and appearance of the towns and countryside in which people live and work. The construction, use, repair, maintenance and demolition of such infrastructure consume resources and energy, and generate waste worldwide. Because of this, construction represents 10% of all economic activity, consumes 40% of the world's material and energy, 17% of its waters and 25% of its timber (41). In the UK the construction industry employs 1.5 million people, consisting of approx 8% of GDP, and annually consumes the equivalent of 6 tonnes of material per head of population (40).

At the end of the last century it was acknowledged that the current performance of construction was unsustainable. The construction industry was subject to a large degree of criticism and regarded as underperforming and having low productivity. *Rethinking Construction* (3) and the UK Governments sustainable construction strategy (42) both emphasised the need for change within the industry.

Sustainability in construction is now a major part of the procurement process, due to global challenges and the development of a business environment where increased levels of transparency are recognised to be best for both business and society. Although there is sufficient evidence to demonstrate a direct correlation between business performance and sustainability, organisations are continuously confronted with diverse barriers to the practical application of sustainability principles. Fundamentally, achieving sustainability requires the process to be reconsidered and a different logic to be applied. The environmental issues are, to a large extent, understood but economic and social impact, and their inter-linkage need also to be taken into account.

For construction to be sustainable, the basic principles of sustainable development must be incorporated. If this can be done, the construction process would be environmentally responsible, have social awareness and be both economic and profitable thereby bringing these objectives to the fore in the built environment for the wider community. The UK Government's strategy for more sustainable construction widens the basic themes to suggest key actions by the construction industry. These include designing for minimum waste, lean construction, minimising energy use in construction, avoiding pollution; preserving and enhancing biodiversity, conserving water resources, respecting people and the local environment, setting targets, and monitoring and reporting to benchmark performance.

21st Century construction needs to evolve from a linear essentially 'physics'-based process, towards non–linear, complex 'biological'-based processes by embracing the advantages provided by modern technology. Such considerations produce the 4Rs cycle – Reduce, Refurbish, Reuse and Recycle – a

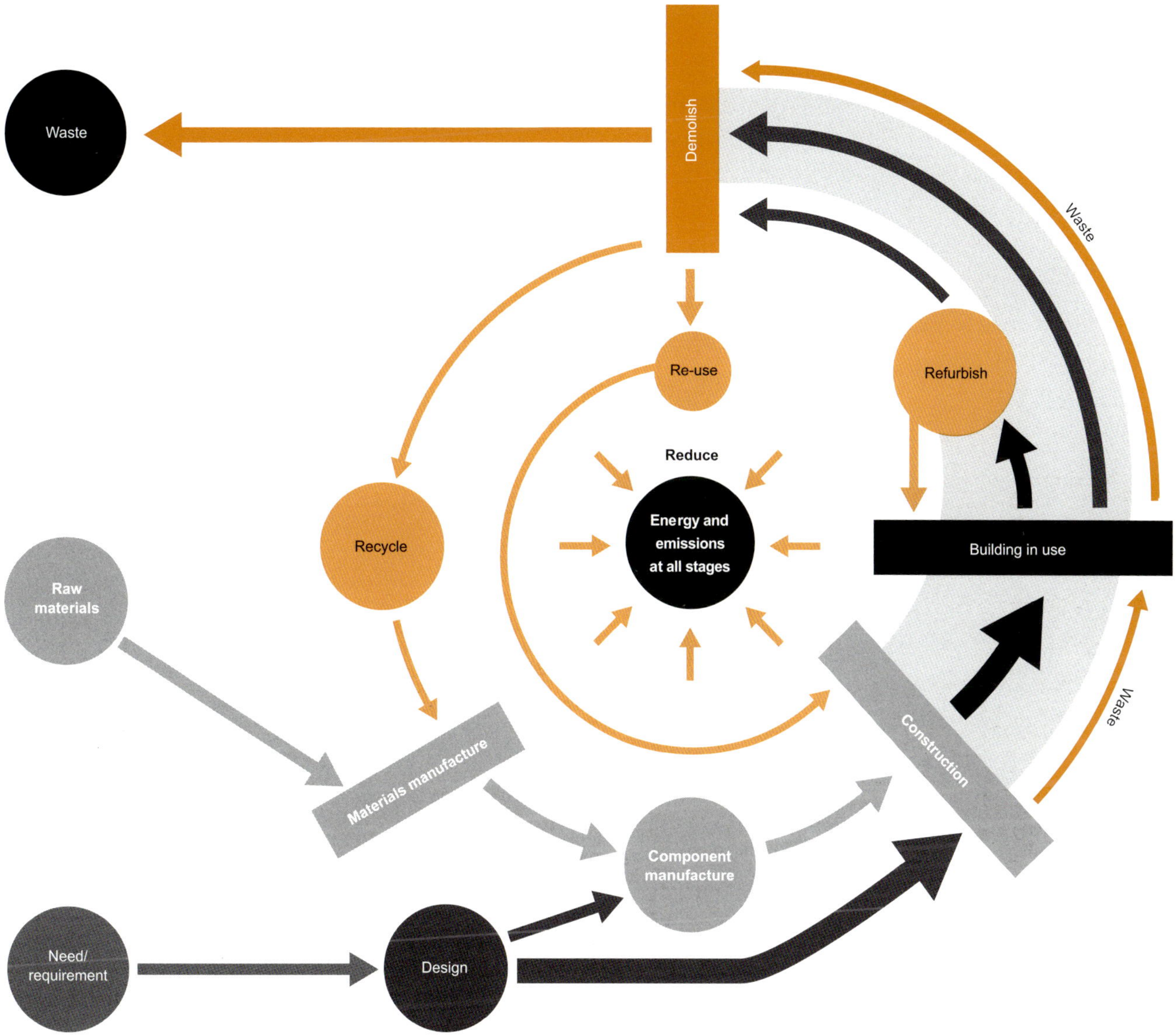

Figure 4.2 The construction cycle process
After Dixon (43)

cycle that echoes biological processes. Adoption of the cycle can benefit the planet by:

- Reducing the emission of greenhouse gases
- Using and reusing resources more efficiently
- Minimising and reusing waste constructively
- Reducing the harmful effects of construction activity.

For construction to be sustainable, it has to provide a flexibly managed 'service' rather than a discrete 'product', with design at the centre of the development process. This concept builds on the Latham Report *Reconstructing the Team* with implementation in the industry requiring a drive for change as predicted by Egan in *Rethinking Construction*. The participants in a project have to appreciate the global consequences of construction activity and be aware of the impact a building has at all stages of its life cycle and understand that, beyond the obvious influences of material and energy use, there are other impacts to consider such as bio-diversity and human health. Therefore, construction project teams need to engage in the multi-disciplinary approach holistically, transparently and collaboratively in order to address the issues relating to sustainability (*Table 4.1*). The procurement process and conceptual design stages of projects are therefore critical because, for something to be and remain sustainable, 'more' is required from 'less' in less total time. In order to achieve this, design has to be recognised as having two overlapping stages:

1. A briefing and concept stage that requires time to develop a finite proposal from the client's brief
2. Design for procurement and safe construction and operation.

One way to facilitate this is to endorse partnering arrangements, as these improve the understanding between project stakeholders and allow agreement to be reached on key issues.

Table 4.1 Issues of sustainable development (43)

Issues	Examples
Appropriate location for development	• Minimisation of environmental and civic costs of getting there • Synergy within towns • Self-sustainability, particularly in rural situations • Synergy of development to existing infrastructure
Choice of built form	• Wall-to-floor ratios, robustness versus recyclability and determinacy • Contribution of building fabric to internal environment by thermal mass • Admittance and control of natural light • Approach to consumption and recycling of water • Flexibility versus adaptability
Choice of structural material	• Control and evaluation of embodied energy/m^2 over the life of a building • Ease of refurbishment • Possibilities for recycling
Minimising the effects of	• Land take, reuse of contaminated construction and deconstruction land • Minimisation of waste and by-products • Ability to reuse materials in future construction

4.3 Sustainable use of timber in construction

When we consider the use of timber, forests provide an 'income' to be used to meet our present needs but need to be 'built' to act as 'capital' for the next generation of planet residents (41). Wood consumes only about 50% of the specific energy (kW/m^3) needed to produce concrete and about 1% of the energy needed to produce steel. This low embodied energy, coupled with the fact that trees can, with good management, improve land quality and soil fertility and are also a prime sink for carbon (carbon fixed from the atmosphere CO_2 by photosynthesis), means timber has the potential to be a completely sustainable resource.

It is well documented that deforestation and forestry practices are of serious concern in the international environmental debate. Sustainable forestry needs to address complex environmental goals such as protection of the forest ecosystem and conservation of biodiversity (abundance of species). It is widely accepted that timber has a lower environmental impact than other construction materials but, due to deforestation concerns, there is increasing pressure to consider its whole life cycle in the form of life cycle assessment (LCA) to verify the actual extent of its environmental impact. The environmental assessment of wood should include harvesting and transport of wood raw material, industrial manufacture of the product, transport to the site of use, building process or installation, and building use including maintenance, demolition and management of wastes.

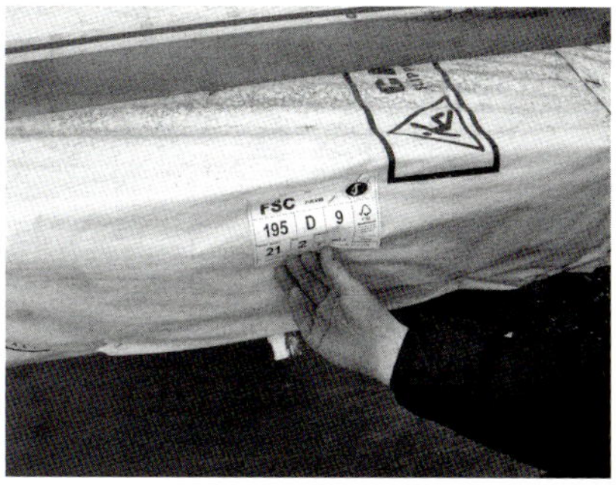

Figure 4.3 Forest Stewardship Council labelling

Various LCA or 'chain of custody' methods include Forest Stewardship Council (FSC) (*Figure 4.3*), Pan European Forest Certification scheme (PEFC), the Canadian Standards Association (CSA) and the American Sustainable Forests Initiative Program (SFI). The chain of custody requires everyone on the supply chain, whether forest owner, harvesting and marketing company, sawmill, or panel and product manufacturer, to be certified in order to ensure timber or timber-based products are procured from well managed sources. Further to this, consideration has to be given to other environmental impacts such as transportation, energy consumption due to process operations and the use of environmentally responsible treatments, for timber preservation.

Large and small scale processes (*Figures 4.4 and 4.5*) can be equally sustainable depending upon the relative demands for market supply. Large scale process will be efficient for supplying large quantities of standardised sawn timber product for mass market consumption. Mass market suppliers will have developed their end product based on utilising the supplied sawn timber section as optimally as possible, thereby reducing waste. Smaller scale operations such as that of MAKAR Ltd, where the timber resource is sourced locally, conversely standardised the sawn section with the end construction method in mind in order to minimise waste. Small scale processes offer the advantage of providing employment to sustain local communities.

The UK Government's policy to improve timber purchasing practices has raised awareness amongst suppliers of the issues surrounding 'legal and

Raw material

Processing of timber

Dry kilning of timber

Storage prior to transport

Figure 4.4 Large scale timber processing for wide distribution

Raw material

Processing of timber

Finished post sections

Built form

Figure 4.5 Small scale local timber processing for local distribution by MAKAR Ltd

sustainable' sources of timber. Increased demand from the public sector has put pressure on the timber trade, and the UK Timber Trade Federation (TTF) has developed a responsible purchasing policy for its members in direct response to the Government's policy. Key to the success of the policy is the Central Point of Expertise on Timber (CPET), which the Government set up to provide guidance on timber from well-managed sources to central Government buyers.

4.4 UK sustainable construction policy

The previous UK government produced initiatives to champion sustainability within the construction industry in order to achieve sustainable development targets. Following on from *Rethinking Construction* [3] the government published two key strategy papers: *A better quality of life – a strategy for sustainable development for the United Kingdom* [42] and *Building a better quality of life – a strategy for more sustainable construction* [44]. The objectives of these strategies (*Table 4.2*) were to promote awareness and understanding of sustainable construction, set out the expectations of Government on construction's contribution to sustainable development, provide information on Government policy for sustainability and stimulate construction which, through the application of lean principle of continuous improvement, would become increasingly more sustainable.

In October 2003 the Better Buildings Summit brought together decision-makers from the house building and construction industries to consider how best to improve the sustainability of all aspects of construction and design, including off-site construction and low carbon technologies. At this summit, the UK Government departments of Defra, DTi (now BERR) and ODPM (now Communities and Local Government) jointly announced the establishment of a Sustainable Buildings Task Group. The aim of the Group was to develop ways in which industry and UK Government can work together to promote sustainable development through better environmental performance in new

Table 4.2 Themes for action [44]

Action	Overview
Re-use existing built assets	Meeting clients' functional requirements may not require new buildings and structures. Refurbishment and/or renovation which improve their sustainability may work better.
Design for minimum waste	Design out waste both during construction and from the useful life – and afterlife – of the building or structure. Think whole life costs. Involve the supply chain. Specify materials with care and consider more efficient use of resources. Think about using recycled materials.
Aim for lean construction	Work on continuous improvement, waste elimination, strong customer focus, value for money, high quality management of projects and supply chains, and improved communications.
Minimise energy in construction	Be aware of the energy consumed in the production and transport of construction products. Adopt 'green' travel policies.
Minimise energy in use	Consider more energy efficient solutions in design including passive systems using natural light, air movement and thermal mass, as well as solutions involving energy produced from renewable sources.
Do not pollute	Understand your environmental impacts and have policies and systems to manage them positively. Use environmental management systems under ISO 14001 or EMAS. Join the Considerate Constructors Scheme.
Preserve and enhance bio-diversity	Look for opportunities to provide and protect habitats throughout the construction process – from the extraction of raw materials, through the construction phase, to the landscaping of buildings and estates.
Conserve water resources	Design for increased water efficiency in building services and water conservation within the built environment.
Respect people and their local environment	Be responsive to the community in planning and undertaking construction. Consider your workforce.
Set targets	Measure and compare your performance with others. Set targets for continuous improvement.

and existing buildings and significantly improve performance on key issues including water, energy, waste and use of certain construction materials such as timber. A main recommendation of the Group was the establishment of a single national code for sustainable building types based on the BRE's Environmental Assessment Model (BREEAM) sustainable construction tool.

BREEAM measures the environmental performance of both design and post-construction stage development. It assesses individual aspects of buildings including management, energy use, health and well being, pollution, transport, land use, ecology, materials and water. This assessment then gives an overall rating of unclassified, pass, good, very good, excellent or outstanding. BREEAM covers a range of building types including BREEAM for schools, offices, retail and industrial units as well as bespoke BREEAM.

Particular to housing was the version of BREEAM known as EcoHomes. EcoHomes provides a rating for new, converted or renovated homes and covers both houses and apartments. EcoHomes balances environmental performance with the need for a high quality of life and a safe and healthy internal environment. The issues assessed are grouped into seven categories: energy, water, pollution, materials, transport, ecology and land use, and health and wellbeing. Many of the issues are optional, ensuring EcoHomes is flexible enough to be tailored to a particular development or market. The

Code for Sustainable Homes (CSH) was developed for the UK Government using EcoHomes as a starting point and complements the system of Energy Performance Certificates for new homes in England and Wales.

CSH has now replaced EcoHomes for all new housing registered in England from April 2007, with a Code rating for new build homes now mandatory, as is the case in Wales and Northern Ireland where a minimum Code Level of 3 is required for new builds. In England, all social housing must achieve a minimum of Code Level 3 or 4 by 2010, Level 5 by 2013 and Level 6 by 2016. The Code for Sustainable Homes has not been adopted in Scotland but a *A Low Carbon Building Standards Strategy For Scotland* has been produced by an expert panel and recommends measures to improve the energy performance of houses and buildings in that country.

Appendix A summarises the Code for Sustainable Homes.

4.5 Case study: Applying the Code for Sustainable Homes using modern methods of timber construction

The CSH is changing the ways in which the fabric of the building is constructed for thermal insulation and air tightness and is affecting the details employed to avoid thermal bridging. Strategies for attaining different Code Levels can be adopted. Up to Code Level 4, either building fabric issues, particularly air tightness, can be addressed, or less attention can be paid to the to the fabric if renewable energy (eg ground source heat pumps) sources can be employed. However, to go beyond Code Level 4 requires both approaches to be considered [(45)].

The first 4 CSH, compliant houses were unveiled in 2007 at the BRE Offsite 07 exhibition, three of which had the majority of their building envelope constructed from timber (*Figure 4.6*). The Sigma Home by Stewart Milne Group achieved Code Level 5 by utilising a closed timber frame system for both the roof and wall; the Ecotech Organics House achieved Code Level 4 by utilising a closed timber frame system for the walls and light gauge steel roof construction; and the Kingspan Lighthouse achieved Code Level 6 by utilising Structural Insulated Panels (SIPs).

Stewart Milne Group Sigma Home

Ecotech Organics House

Kingspan Offsite Lighthouse

Figure 4.6 BRE Offsite 07 exhibition

One of the main aims of the event was to demonstrate the capability of off-site MMC in achieving higher performance levels than more conventional construction forms. Employing MMC techniques improved the construction time for all methods, for example reducing the construction time of the Sigma Home to eight weeks relative to the conventional 24 weeks required for two semi-detached four-storey structures (note: the house is the height of a large three-storey, pitched roof property, but is arranged over four levels to minimise the building footprint and maximise affordability). Notable from the Lighthouse construction process was the fact that built form can impinge upon construction efficiency with the curved roof form proving problematic to erect and seal [46]. Another advantage of MMC was the use of robust, simple and repeatable details; these improved the air tightness and acoustic performance of the systems. Airtightness was particularly noticeable in the Ecotech House where the design for $3.00m^3/h.m^2@50Pa$ was improved to 1.14 and $1.94m^3/h.m^2@50Pa$ for the flat and house respectively as demonstrated by precompletion testing.

Sigma Home pod installation

Ecotech closed panel erection

Lighthouse SIP panel

Figure 4.7 Use of off-site MMC to deliver Code for Sustainable Homes requirements
Copyright BRE, reproduced with permission [45]

In modern construction, apartment blocks and terraces in particular, acoustic transfer over party walls is an issue. The SIGMA Home was the only attached property and to achieve the party wall requirements it used Robust Details (www.robustdetails.com). Robust Details Limited provides a route to compliance for Part E of the Building Regulations (England and Wales) and the detail utilised in the SIGMA Home achieved 5dB above requirements to provide an additional 5 credits (other details exist to provide 8dB, the improvement required for 4 credits, but, these Robust Details did not exist at the time of construction). The detached houses achieved the full 4 credits for sound performance.

The nature of the timber construction forms adopted for the Ecotech, Lighthouse and Sigma constructions are lightweight, have a high degree of thermal insulation and are built to be airtight. Because of this, there is a risk of overheating due to a lack of thermal mass. Overheating can be caused by either external sources (ambient hot air, solar radiation etc) or internal sources (cooking, electrical equipment, the occupants etc). To overcome these problems, the designs incorporated low energy electrical goods and lighting, insulated hot-water systems and were well designed for shading, solar gain and ventilation. Further to this, both Kingspan Off-site and Stewart Milne Group used the project as a live experiment to evaluate the performance of phase-change materials. 'Micronal' from BASF was trialled in the Lighthouse and 'Energain' from Dupont in the Sigma design. Both of these systems use a wax that melts at 26°C and acts as a form of latent

Table 4.3 Building fabric performance

Design Code Level	Ecotech*		Stewart Milne Group Sigma	Kingspan Offsite Lighthouse
	Flat	House		
	4	4	5	6
U-values (W/m²K)				
Doors	1.00	1.00	1.20	0.31
Windows	1.30	1.30	0.68	0.68
Roof Windows				1.40
Ground floor	0.11	N/A	0.18	0.11
Walls	0.15	0.15	0.15	0.11
Main roof area	N/A	0.06	0.15	0.11
Other roof area	N/A	N/A		0.05
Airtightness (m³/hm²@50Pa)				
Designed	3.00	3.00	1.00	1.00
Achieved	1.14	1.94	2.30	1.27
Heat loss parameter (W/m²K)				
Designed	0.7	0.78	1.10	0.80
Achieved				0.80

*The Ecotech Organics House consists of a 2-storey house on top of a 1-storey flat

heat storage that absorbs heat from the building as it starts to overheat. The application of plasterboard as a building performance regulation requirement may impinge upon the benefits of such systems, but post-occupancy testing will act as a method of determining their effectiveness.

Both the Lighthouse and the Sigma Home incorporated micro-renewable technologies needed to achieve a Code Levels 5 and 6. The roof of the Lighthouse was designed to take the requisite 47m^2 of photovoltaics needed to meet the electricity needs (a Code Level 6 requirement). The Sigma Home adopted a combination of both photovoltaics and wind turbines. However the turbines achieved a low output due to relatively low local wind speed. Consequently, Stewart Milne fitted more photovoltaic panels to make up the deficit.

4.6 Case Study: Applying the Passive House concept using modern methods of timber construction

In many parts of Europe, primarily Germany, Austria and Switzerland, there is an increasing move toward the use of the Passive House Standard (from the German Passivhaus). Passive House is a holistic approach intended to improve the building envelope to a degree allowing for substantial simplification of the heating system. By improving air tightness and thermal insulation, together with supplementary supply air heating (mechanical ventilation with heat recovery), Passive Houses achieve a comfortable indoor climate in both summer and winter without needing a conventional heat distribution system. To permit this, it is essential that the building's heating load does not exceed 10 W/m^2, a level roughly equivalent to an annual space heat requirement of 15 kWh/m^2. Passive Houses thus need only about 20% of the space heating in more conventionally designed houses.

A key driver underlying the Passive House concept was that the design of buildings required substantial rethinking. Previously, the energy efficiency of buildings was increased by improving the component parts and incorporating renewable technologies whilst ensuring the building envelop was airtight. However, it was recognised that these historical attempts to improve efficiency added to cost and impinged upon indoor air quality. In many respects the houses showcased at the BRE Offsite 07 project have exhibited the same findings, with the micro-renewable technologies employed failing to provide the return period or efficiency relative to cost necessary to substantiate endorsement.

Passive House works on optimising energy-saving features (*Table 4.4*). The simplistic heating system fulfils two criteria:

1. Thermal comfort: the heat distribution system is simplified by maintaining the surface temperature of the outer wall and windows close to the room air temperature. This allows for thermal comfort without the need to place radiators at outer walls, especially below windows, for the compensation of radiative asymmetries and cold air downdraught.
2. Space heating: the space heat demand is kept low enough in order that space heating can be provided by the ventilation system alone, at hygienic flow rates, without the need for recirculation or for any additional water-based heat distribution system. This allows for very simple and cost effective air heating systems. Problems frequently associated with air heating do not occur provided that the requirements of the Passive House standard are met.

Further energy demands for household appliances can be met by renewable sources, thus giving the Passive House a net zero carbon impact. The use of

Table 4.4 Basic features of Passive House construction

Features	Examples
Compact form and good insulation	All components of the exterior shell of the house are insulated to achieve a U-value that does not exceed 0.15 W/m²K.
Southern orientation and shade considerations	Passive use of solar energy is a significant factor in passive house design.
Energy-efficient window glazing and frames	Windows (glazing and frames, combined) should have U-factors not exceeding 0.80 W/m²K, with solar heat-gain coefficients around 50%.
Building envelope air-tightness	Air leakage through unsealed joints must be less than 0.6 times the house volume per hour at 50pa (6-10 times better than standard UK construction). To achieve this air tightness foil is applied for all joints of building components.
Passive preheating of fresh air	Fresh air may be brought into the house through underground ducts that exchange heat with the soil. This preheats fresh air to a temperature above 5°C, even on cold winter days.
Highly efficient heat recovery from exhaust air using an air-to-air heat exchanger	Most of the perceptible heat in the exhaust air is transferred to the incoming fresh air (heat recovery rate over 80%).
Hot water supply using regenerative energy sources	Solar collectors or heat pumps provide energy for hot water.
Energy-saving household appliances	Low energy refrigerators, stoves, freezers, lamps, washers, dryers, etc. are indispensable in a Passive House.

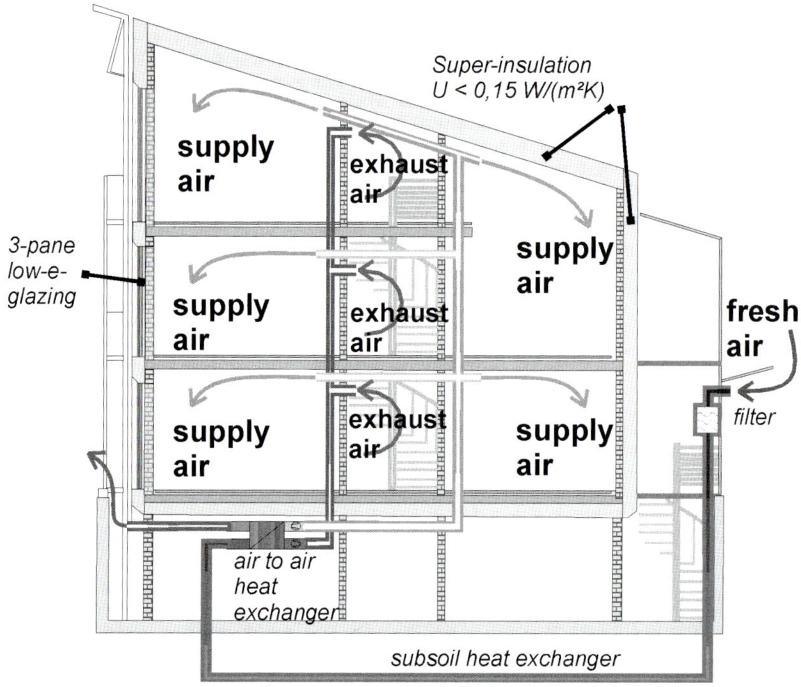

Figure 4.8 Schematic of the Passive House systems
After Passive House Institute

these concepts has been demonstrated by the Cost Efficient Passive Houses as European Standards (CEPHEUS) project. This tested their viability at a European level on a total of 221 housing units in 14 building projects. The projects covered a range of building types (commercial and domestic) and were constructed using a range of methods, two of which had the building envelope fully constructed from timber (*Table 4.5*); other projects often employed a hybrid system. These projects are now occupied and successfully met functional requirements and space heat savings targets.

Passive House has been demonstrated primalily in continental Europe with more than 7,000 dwellings constructed, although there are examples in the UK. Elgin-based RTC Timber Systems Ltd has developed a closed panel timber system capable of meeting Passive House standards, which uses solely local Scottish products for the building envelope (*Figure 4.9*).

Table 4.5 Examples of timber Passive House construction in Europe [47]

CEPHEUS Project	U-values (W/m²K)		Heating	Ventilation	Hot water
Sweden, Gothenburg – Project 03	External wall	0.08	Direct electric supplementary heating of supply air	Balanced supply and extract air flows with heat recovery from extract air	Solar collectors (5 m²) and 500 litre storage per dwelling unit
	Ground floor	0.09			
	Roof	0.07			
	Whole window	0.88			
Switzerland, Nebikon – Project 13	External wall	0.11	Preheating of fresh air by subsoil heat exchanger, packaged ventilation unit with air-air/water heat pump for supplementary supply air heating and hot water heating	Controlled ventilation supply and extraction with heat recovery from extract air	Service water storage is heated by the heat pump of the packaged ventilation unit or directly electrically
	Ground floor	0.11			
	Roof	0.11			
	Whole window	0.88			

Considering Passive House relative to the CSH, a well designed Passive Energy House will achieve the energy credits needed to achieve Code Level 4 and, with minimal uplift, can be improved to reach Code Level 5. The energy modelling tool for CSH is SAP, but to design a passive energy house a more sophisticated tool needs to be applied, such as the Passive House Planning Package (PHPP) which is sensitive to location, overshading and orientation and, as a result, is considered a more reliable indicator of the actual performance of the building. SAP was created as a 'compliance tool' whilst PHPP is used by designers from 'concept development' through to 'detailed design', ensuring that performance targets are met without stifling architectural expression [48].

Figure 4.9 RTC Timber Systems Passive House under construction in the UK

5 Design form and function

5.1 Timber properties

Wood is a natural, heterogeneous, anisotropic, hygroscopic composite material. From growth to use its structural properties are highly variable as a result of a whole range of influencing factors. The effects these influencing factors have on the structural properties of the timber section proposed require careful consideration. If a factor is deemed to be negligible in the overall scale of investigation, then it can be ignored.

The structure of timber can be considered at four levels:

1. Micro: Cell level
2. Messo: Growth ring level
3. Macro: Clear wood level
4. Massive: Sawn timber

Although an appreciation of the four levels is beneficial when considering timber as a building material, it is the "massive" structure that is most relevant. In this context massive wood means large dimension timber produced by sawing, with the smallest cross-sectional dimension generally being more than 100mm, or a combination of solid wood members created by glueing together relatively small pieces of lumber (for example, glulam beams and cross-laminated timber panels). It is massive wood that is used in structural design and, because of the inherent presence of imperfections, structural wood has lower strength than clear wood of the same species. Therefore, when designing with timber it is important to have an appreciation of the overall properties of structural timber.

Appendix B is an overview of timber design properties.

5.2 Structural timber design

5.2.1 Code of Practice

UK structural timber design practice was, until April 2010, in accordance with BS 5268-2:2002, a code based on permissible stresses [(49)]. Permissible (working) stress has been the basis for the UK's formal national timber codes for approximately 50 years and it is one in which safety factors are already incorporated into the tabulated material properties. Adoption of the Eurocode Suite has standardised design procedures across Europe. Eurocode is a limit state design code of practice; limit state codes link the structural reliability to clearly defined states beyond which the structure no longer satisfies specified performance criteria.

The transition from current national codes of practice to one that covers all EU member states has been a gradual implementation of change known as the harmonisation period. The main objective of Eurocode is to further facilitate the free trade of construction products and services within Europe. Design of Timber Structures will be in accordance with Eurocode 5 [(50)] which is to be used in conjunction with Eurocode 0 [(51)] Basis of Structural Design and Eurocode 1 [(52)] Actions on Structures. Eurocode is based on two types of limit state: ultimate limit state and serviceability limit state (*Figure 5.1*).

Instance where serviceability limit state has been breached

Instance where ultimate limit state has been reached

Figure 5.1 Examples of instances where limit states have been breached

Timber cladding attached to SIPs

Brick slips

Figure 5.2 Alternative cladding methods

Ultimate limit states are those associated with collapse or with other forms of structural failure. Ultimate limit states include: loss of equilibrium; failure through excessive deformations, transformation of the structure into a mechanism, rupture and loss of stability.

Serviceability limit states include: deformations that affect the appearance or the effective use of the structure, vibrations which cause discomfort to people or damage to the structure, and damage (including cracking) that is likely to have an adverse effect on the durability of the structure.

Limit state design presents the designer with more scope for design input. Current practice of permissible stress design is based on factored characteristic strength values. The incorporation of safety factors prior to actual design work limits the engineer's input based on the particular design circumstance. An example of this is how the hygroscopic nature of timber is accounted for in Eurocode by assigning a service class that depends on the moisture content of the surrounding atmosphere and its expected fluctuations (*Table 5.1*) and then factoring the characteristic material values accordingly. The permissible stress design procedure was more prescriptive as, although it accounted for service class conditions, it did not facilitate the same level of engineering input. Allowing the designer to consider the service class for the design, as appropriate, will correlate the level of safety factor required for the given circumstance.

Table 5.1 Definition of service classes and examples [53]

Service class	Temperature	Approximate maximum humidity	EMC* %	Examples from the EC5 National Annex
1	20°C	65%	12%	Warm roofs, intermediate floors, timber-frame walls – internal and party walls
2	20°C	85%	20%	Cold roofs, ground floors, timber-frame walls – external walls, external uses where member is protected from direct wetting
3	Conditions leading to higher moisture contents than service class 2.		> 20%	External uses – fully exposed
*EMC = Maximum equilibrium moisture content for most softwoods.				

In theory the Eurocode should result in timber design that is economic, serviceable and ultimately safer. Eurocode 5 facilitates a wider selection of materials and components, and also provides more guidance on the design of built-up components than BS 5268 Part 2. This will facilitate the incorporation in design of new engineered products and allow future products to be integrated for use.

5.2.2 Design responsibility

Presuming a building is properly constructed, maintained and used as intended, the responsibility for ensuring that neither the ultimate or serviceability limit states are infringed lies with the engineer. To ensure the overall durability of the system the engineer has a duty of care to ensure that architectural detailing is suitable and, where necessary, preservative or protective treatments for timber materials and metal fastenings are specified. Primarily concerned with the adequate specification of load-bearing members, components and assemblies, the engineer also has to consider non load-bearing items where the integrity of their fixing has safety implications, eg the attachment of external cladding (*Figure 5.2*). When considering off-site MMC products, it is also important to assess the transportation and construction methods since the actions applied during these processes will be different from those that are applied in service. Because of this, further design considerations need to be taken in order that the system is suitably robust.

5.2.3 Impact of standardisation

Standardisation of components or parts can result in a degree of overspecification that has to be balanced with overall efficiency gains in production and construction (*Figure 5.3*). As an example, imposed loads are dependent on the designated use and geographical location of a building and this influences member sizing, the need to design for disproportionate collapse and the need to specify corrosion and protective treatments. In the case of standardised system components, allowance will have to be made for this. It may also be a requirement to standardise for the worst case scenario or to designate different standard components for different geographical location or design scenarios.

5.2.4 Design life

The building must be designed to have adequate strength, stability and structural serviceability during construction (the execution phase), throughout its design life and, in the case of accidental design situations, the system must be designed in order that it is suitably robust.

The structural adequacy of a system has to be considered during the transportation and construction phases (*Figure 5.4*). The transportation and construction of factory-produced components may result in different applied forces than will be experienced in service and these must be taken into account.

In most design circumstance it is common for the design life to be specified as 50 years. However, a properly designed and maintained timber building can last for centuries. For less permanent structures a 10-year design life may be assumed and it may also be considered that some elements of a structure have a shorter design life and will be replaced as part of the maintenance management. For example, a component may be specified purely to provide system rigidity during transportation or for lifting operations during construction.

Sacrificial layer of OSB to act as temporary working platform during construction to improve efficiency of construction process

SIP roof lifting operation applying force different to those which will be applied after construction

Figure 5.3 Applied actions different to in-service conditions

Panaloc trailer for vertical storage of wall panels and safe working access

SIP roof section being craned into position

Figure 5.4 Designing for transportation and erection of manufactured elements

Temporary bracing during construction for stability

Holding down detail

Figure 5.5 Temporary and permanent methods of providing resistance to applied wind action

Timber structures are relatively light in weight and, as a result, checking for stability is important. Wind force, both during the service life and construction or execution phase of a structure, requires adequate resistance to uplift, sliding and overturning to be specified (*Figure 5.5*).

Once constructed timber buildings may rely, depending on the number of storeys, on the weight of the roof to resist overturning or sliding. Similarly the resistance of timber frame buildings to horizontal wind forces may depend on the connection of the roof to an adjoining terrace of houses, or on the shielding effect of brick cladding, Because of this, temporary bracing may be needed during the execution phase.

The Construction (Design and Management) Regulations 1994 (CDM) [54] require the engineer to consider and, where necessary, to provide safe procedures or instructions for:

- the method of erection
- the type of craneage
- lifting points
- manual handling.

The engineer should ensure that this information is recorded in the Health and Safety Plan and the Health and Safety File.

Where design assumptions dictate certain methods and sequences of erection, full information concerning them should be included in the Health and Safety Plan as well as being shown on the design drawings.

5.3 System deformation

Failure to make provision for movement in the design of timber buildings can cause serious serviceability problems. The overall height of a timber building reduces with time, due to:

- shrinkage of the timber as it dries out
- compression due to the building's self weight and occupancy
- settlement at the interfaces of roofs, walls and floors under load.

System movement will also occur as a result of connection slippage and creep effects.

5.3.1 Timber movement

Being hygroscopic, the moisture content of timber varies with relative temperature and humidity; variation results in radial and tangential movements. Along the grain, however, it is relatively stable. Movement varies depending on the species but tangential movement is typically twice that in the radial direction. Changes in dimension are therefore not equal in all three principal axes and this is further complicated by timber cutting practices that have an effect on grain orientation. The average of the tangential and radial movement should therefore be taken into account at the design stage. Also, due consideration should be taken at the design stage to the nature of the product being specified, whether it is kiln-dried structural graded softwood or an engineered product.

Kiln-dried structural graded softwood normally has a moisture content of 18-20% on delivery and will dry out in a heated environment to about 10-15%, resulting in an across-the-grain dimensional decrease of 1-2%. If dimensional movement is critical, 'super-dried' solid timber can be specified with a moisture content of 14% or less and a moisture resistant coating that colours timber blue for identification.

Engineered wood products normally have a lower moisture content and are consequently less susceptible to shrinkage. Although they are generally more expensive, they can also be considered when trying to reduce movement.

Sound construction detailing and design, in conjunction with quality fabrication and execution, will restrict settlement at member interfaces. Quality assurance of the products used in fabrication procedure will restrict natural wind or twist of horizontal members. These flatten out under load and can lead to movements of up to 3mm per storey when the complete roof has been installed.

Further to this, induced stresses result in elastic and creep compression, the extent depending upon the dead and imposed load and properties of the timber product specified.

5.3.2 Differential movement

It is normal for the overall movement of timber in a structural system to be negative due to a combination of shrinkage, settlement and elastic and creep compression. However, it is normal for structural systems to be made of a combination of materials that act differently, in different environmental conditions, and it is therefore important that design and detailing take this into account.

The coefficient of thermal expansion of timber along the grain is smaller than that of steel or concrete. In hotter environments any expansion across the grain due to temperature is more than counteracted by shrinkage due to the reduced water content. Also timber's thermal conductivity is much lower than that of steel, so it is less responsive to ambient changes in temperature than steel. Changes in temperature can therefore cause short-term differential movement, because timber and wood-based materials have lower thermal expansion characteristics than those of masonry and steel.

Claddings that are clay-based products absorb moisture from the atmosphere after construction until equilibrium is reached and this results in expansion. The level of movement varies depending upon the type of clay used in the production process, but is typically 0.25-1.0mm per metre of height.

Movement tolerance can be built into the system at the design stage by means of movement joints or flexible connections such as proprietary lintels, separate sills and specially designed wall ties.

Sequence of construction can also alleviate differential movement. The timber load bearing systems can be fully erected and loaded as much as practically possible (ie the roof can be clad) prior to the external cladding being installed. This minimises relative movement between the timber system and the cladding due to compression and settlement. However, where cladding is installed in the factory, due allowance needs to be made to accommodate any relative movement.

5.3.1 Joint slippage and tolerances

The transfer of forces and final deformation of a timber structural system is affected by local displacement of the connections, known as joint slip (*Figure 5.6*). Depending upon the structural system and methods of fabrication being employed, a variety of connection types will be used ranging from glued joints, which experience little or no slip, to nails which experience a high degree of slip.

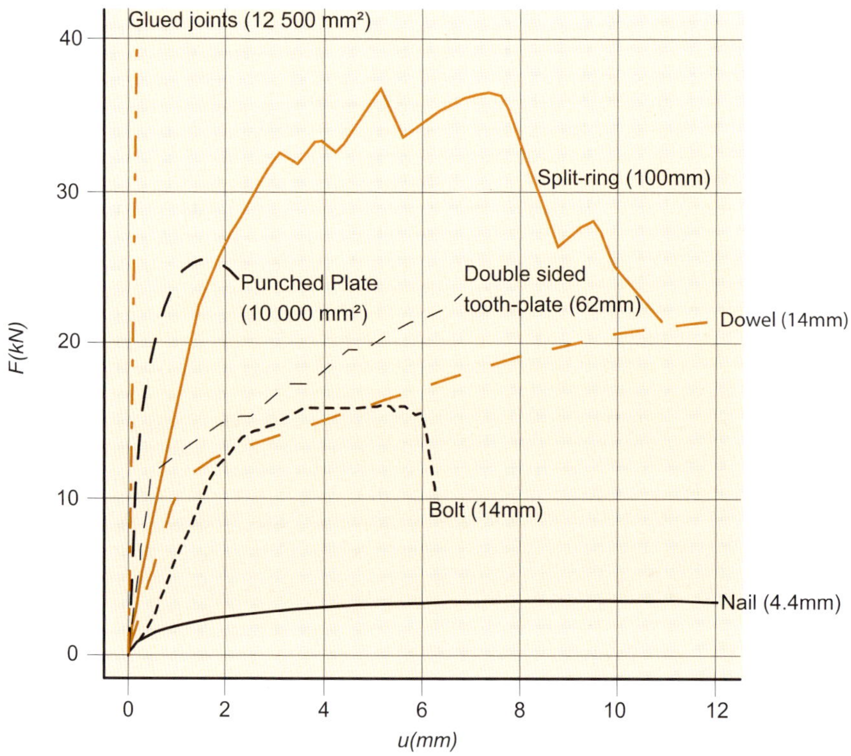

Figure 5.6 Experimental load slip curves for joints in tension parallel to the grain (86)
After Racher

Connection deformation can be determined either by conducting tests in accordance with the appropriate Code of Practice or by determining the overall performance of the joint from the behaviour of the individual fasteners. It is also important in design to consider tolerances, especially when bolted connections are used, as over-sized bolt holes will result in an initial slip taking place.

Tolerance is an important design consideration when building with an engineered product. The higher the level of prefabrication carried out off-site, the tighter building tolerances need to be on-site. If building tolerances are poor, then the system will not fit together as intended and alterations late in the design stage are difficult and costly. Foundation tolerances are also critical. If the foundation level is out by a large margin, then the prefabricated, often rigid elements, will not connect together appropriately on-site. Poor tolerances can impact on building performance parameters such as airtightness and structural robustness as a result of reduced connectivity, continuity and tying in.

5.3.2 Creep
Because it is a viscoelastic material, timber experiences deformation over time under a constant applied action known as creep. The parameters influencing creep in wood are load duration, moisture content, temperature and stress level. Interaction occurs between all the parameters, but only the combined effects of load duration and moisture content are taken into consideration in design rules [55].

5.4 System robustness and progressive collapse
During the design life of a building, accidental actions can take place that impose loadings on the structural system but were not originally designed for, for example exceptional snow drift, impact, fire and explosion. Although these actions are unlikely to occur within the design life of the structure, if they do occur, then the system will be suitably robust to ensure that the consequences of the action will not be disproportionate to the cause. To provide additional robustness, redundancy can be introduced into the system via additional tying in order that a redistribution of forces within the system can occur should an accidental action take place (*Figure 5.7*). Panellised buildings for instance can be susceptible to progressive collapse due to a lack of redundancy; the loss of one component redistributes load or adds debris loading and can lead to the sequential failure of other elements.

Ring beam over lintel providing redundancy to system

Tying of system together

Figure 5.7 Methods of introducing system robustness

Overall system performance is as a result of the combination of building parts (roof, walls and floors). The way they are connected together will affect the overall performance of the system. As an example, imposed loads are dependent on the designated use and geographical location of a building and this influences member sizing, the need to design for disproportionate collapse and to specify corrosion and protective treatments.

In the case of standardised system components, a generic specification is needed. It may be a requirement to standardise for the worst case scenario or to designate different standard components for different geographical locations or design scenarios. The design might also include 'sacrificial' materials to improve the build process, for example, an additional layer of OSB to form a working platform from floor cassettes during construction, prior to installing the chipboard flooring once the building is windproof and watertight. The inclusion of sacrificial materials can add to the structural performance, for example, the additional layer of OSB if properly fixed can increase the rigidity of the floor cassette. However, it will also add to the permanent action on the system once erected and may, therefore, be important in longer spans.

Partnering facilitates a better understanding between the designer and the client about the building construction process and building use. There is a clear requirement to design for the likelihood of specific occurrences and corresponding magnitude of accidental actions throughout the design life of the system. The designer has to consider the construction procedure (for example, the constructor's ability to make the designed connections properly) and make adequate provision for access to show routine maintenance and inspection can take place. The design must also allow for applied actions during the construction process due to level of prefabrication and execution sequence. The CDM Regulations require the Engineer to consider

Figure 5.8 Storage of internal partitions

risks to the health and safety of those engaged in the maintenance of the structure, to provide safe means of access to elements needing maintenance, and to ensure that this information is recorded in the Health and Safety Plan and the Health and Safety File.

Durability is important and the building should be designed, manufactured and constructed in a manner that ensures it will satisfy all the relevant limit state requirements for its specified design life. It is important, therefore, to consider the nature and behaviour of timber and how it interacts with other materials. For example, the timber must remain dry and well-ventilated and, where this cannot be ensured, a durable species or suitable protection treatment should be specified (*Figure 5.8*).

5.5 Fire resistance

Depending on the nature of the building, the built off-site MMC timber system will have to conform to the relevant Building Regulations. In relation to timber systems, it can also be important to consider the management of on-site activities to ensure fire does not occur during the construction phase. However, the more work carried out in the factory, the more protected from fire the system will generally be as it will be nearer its completed state, that is the plasterboard will have been installed (*Figure 5.9*). Also, active fire prevention measures such as sprinklers can be installed in the factory and, as a result, can be put into service more quickly.

Preinstalled sanitary ware

CNC fabrication

Figure 5.9 Bathroom pod with preinstalled services and plasterboard providing fire protection and facilitating speed of construction

Depending on their function, members, components and assemblies exposed to fire may be required to provide:

- mechanical resistance (to remain sufficiently strong and stiff to permit a safe exit from the building)
- integrity (to form an effective barrier to smoke and flame)
- insulation (to limit the transfer of heat by conduction from one area of fire to another) or a combination of these.

It may also be necessary to limit the thermal radiation to the side unexposed to fire and to ensure that the surface spread of flames does not unduly endanger the occupants of the building or neighbouring buildings.

To ensure adequate mechanical resistance of structural assemblies and members in fire, there are two main approaches:

- insulating them from the heat
- specifying sacrificial timber.

Structural elements including mechanical fasteners can be insulated from heat by covering them with one or more layers of insulating material of a specified thickness, such as plasterboard. Specifying sacrificial timber involves checking that the dimensions of the residual sections of timber members are, after charring, sufficiently large to support the design loads with acceptable deflections within the specified period of fire resistance. Where necessary the section size can be increased to provide the required period of fire resistance. This approach can be used for exposed columns or beams and is applicable to solid timber, glulam, LVL and other structural timber composites, but not to engineered timber joists.

6 Re-engineering timber

6.1 The need for re-engineering

The uses for timber have changed over the centuries. In its natural form, timber has its limitations due to availability, specifics of serviceability and dimensional restraint. The properties of timber vary from species to species, are dependent on the growth environment of the species and also vary across the structure of the species itself. Timbers with a cross-section of greater than 75x225mm and longer than 5m, and which can be used for structural purposes are at a cost premium. It may therefore be possible to specify a tropical hardwood for a specific design circumstance but the cost and availability make it unfeasible. The re-engineering of timber, in the form of timber composites and timber engineered products, has challenged these restrictions.

6.2 Timber composites

There is evidence of timber composites dating back to the Egyptian Pharaohs; archaeologists have found traces of laminated wood in their tombs. Plywood, however, is widely regarded as the original engineered timber. The English and French are reported to have worked wood on the general principle of plywood in the 17th and 18th centuries but the industry was, according to the American Plywood Association (2007), 'born' in 1905 when Gustav Carlson decided to laminate wood panels from a variety of Pacific Northwest softwoods called '3-ply veneer work'. A further significant advance in plywood occurred in 1934 when Dr. James Nevin, a chemist at Harbor Plywood Corporation in Aberdeen, Washington state, finally developed a fully waterproof adhesive and this subsequently opened up new markets for the product.

It is the concept of plywood that forms the basis of most engineered timber composites manufactured from timber sections and reconstituted timber through adhesion. Timber composites overcome the dimensional limitations of sawn timber and improve performance, structural properties and stability. They transform the natural orthotropic product into one with more homogenous properties and optimise the use of a valuable resource whilst minimising waste [(56)]. Examples of timber composites that are used extensively in structural applications are glue laminated timber (glulam), laminated veneer timber (LVL), laminated strand lumber (LSL), parallel strand timber (PSL), cross laminated timber (CLT) and oriented strand board (OSB). Depending on the country or region of use, these products must comply with standards in order that they can be specified in design by the engineer.

6.3 Massive timber

Countries such as Austria have a surfeit of timber sourced from sustainably managed smallholdings, making it efficient to use more, rather than less, material when secondary considerations such as insulation, acoustics or dynamics, merit heavier sections [(57)]. As a result, massive timber is manufactured and more readily employed. Massive timber is normally a bi-axial or cross-laminated timber that is manufactured from low grade timber off-cuts that have been reduced to strips and glued in perpendicular layers under high pressure, either mechanically pressed or in a vacuum bag. The cross-laminated solid timber (CLT) panels can be cut in the factory to the required shape and size. Further to this, routing and holes for services and connections can also be incorporated within the panels.

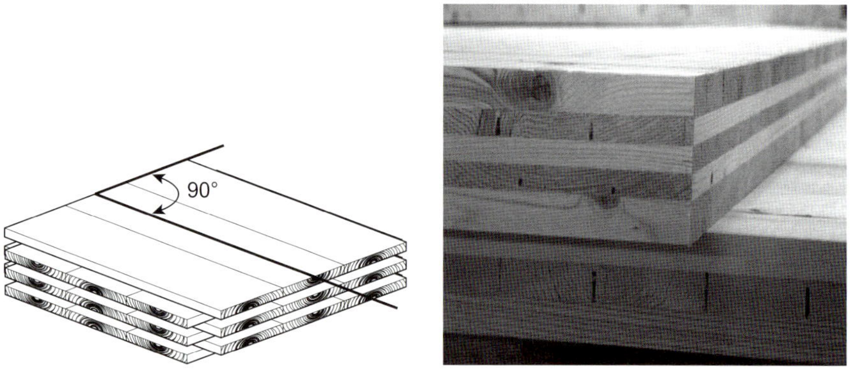

Figure 6.1 Cross laminated structure of solid timber panels
Photo: Finished product by Finnforest

Orientation of laminates (58)

Finished product (59)

Massive timber can be used for roof, wall and floor construction. Structurally, panels have high in-plane stiffness and will span in both directions. The cross-laminated products are also more stable, and swelling and shrinkage are minimal. At the construction stage, the panels are connected together with mechanical fixings at regular intervals to avoid concentrated loads and to distribute the connection stresses to avoid the risk of overloading any one connection which might result in unzipping failures (58). The case study in Section 6.7 illustrates CLT panel construction.

Timber frame construction is more limited when considering multi-storey construction beyond four storeys because it becomes more difficult to achieve stability, fire protection requirements become more onerous and it is harder to accommodate movement. This said, five- and six-storey timber frame systems have been built and, in the UK, timber frame can go up to 18 metres within the building regulations. Guidance for multi-storey timber frame construction is available from TRADA Technology in a design guide produced with BRE (59).

However, massive timber systems are inherently more rigid and the solid cross section can be increased to provided additional fire protection through charring (CLT panels have a charring rate comparable to softwood of 0.7 mm/min). In relation to movement the comparative frame shrinkage of a CLT structure, due to change in moisture content, is much less than in a conventional timber frame structure. This said the effects of creep become significant above about 12 storeys and must be considered in the detailing of claddings, linings and vertical runs of building services (60).

From a thermal performance perspective cross-laminated timer can also perform well. For example, a standard 100mm cross-laminated panel of density 500kg/m^3 has a U value of 0.13 W/mK. Massive timber systems also provide thermal mass that can improve design if passive heating techniques are employed. The thermal mass of cross-laminated timber will store energy during the day from heat generated in the building as a result of usage or passive solar gain and this energy is subsequently released later in the cycle.

Of course, thermal performance is compromised if the construction is not airtight and, in order to achieve this, joints need to be detailed and constructed properly. This is more critical in CLT structures because a vapour control layer is not standard. Because of this and with joints being normally screwed together, either pre-compressed foam tape is applied within the joint or breathable tape is applied across the outside, since the airtightness would otherwise be reliant on 'true' surfaces being achieved. Detailing is also important when considering acoustic performance. Specifically, party wall details need to be robust in order that flanking does not take place. To achieve the Building Regulation requirement, the acoustic resistance of the panel can be taken into consideration (this will depend on density and thickness) and plasterboard will also normally be specified to achieve the required acoustic performance.

6.4 Timber trusses

Timber trusses were the first engineered product that allowed timber structures to span beyond the limitations of 5 – 7m. The Romans were first to develop triangulated trusses with spans up to 30m for the roofs of their basilicas and these greatly influenced the form of medieval Italian and later European roof structures [61]. Trusses produced from structural timber are still the most commonly used method of roof construction in domestic dwellings. Modern truss systems are manufactured from sawn timber sections joined together at the node points by pressed metal plate connectors. Software packages are used to design the full roof system and to provide the information for cutting the timber and specifying the metal plates based on the load/span conditions of the particular design case.

6.5 I-beams

A concept to increase the structural performance of timber by means of optimising the cross section (originally explored by Victorian engineers to increase the structural performance of steel and iron by producing I- and box beam sections) was first researched between 1915 and 1942 in both North America and Europe. Subsequently timber I- and box-beam sections with plywood webs and solid timber flanges were widely used for aircraft and glider wing spars and struts.

Again the catalyst for mainstream use of timber in the form of an engineered product was the need to build rapidly. After WWII, the requirement for schools, community centres and telephone exchanges was satisfied, with substantial volumes of industrialised building elements being produced each week by several British factory lines. Many such elements remain satisfactorily in use to this day.

A modern I-joist (*Figure 6.2*) consists of LVL or high-strength-graded timber flanges and oriented strand board (OSB) or timber or metal truss webs (posi-joists). Timber I-beams are now extensively used for floor construction in the UK due to the advantages they offer over solid timber sections in terms of their span capabilities, light weight and ease of design and construction. A quality assured product, I-joists are manufactured in controlled environments and must be adequately protected when stored.

Major I-beam producers include Weyerhaeuser (USA), Finnforest (Finland) and James Jones (Scotland), all of whom supply the components for full

OSB web

Timber truss web

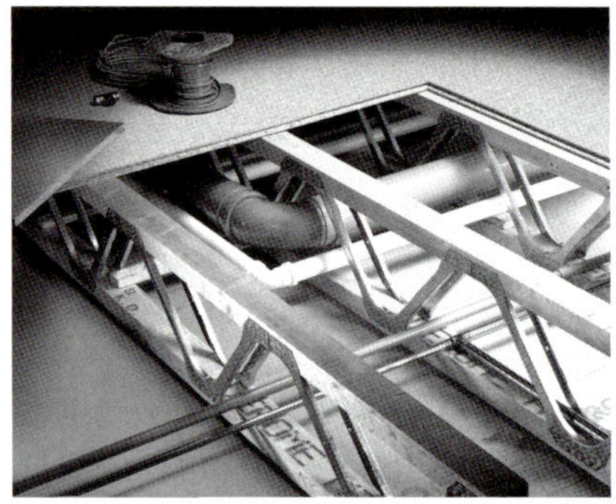

Steel truss web

Figure 6.2 Examples of timber I-joist flooring systems
Photo steel truss web: MiTek

floor systems using their specified products and software. Should the products and software of the company supplying the flooring system not be used in combination, the floor will not be guaranteed by the joist supplier. Floor cassette systems were traditionally formed from solid section timber. These are now normally manufactured using a combination of I-joists and timber composites. Solid section timber floor cassettes are still used today in situations where load/span conditions are less onerous. However, the I-joist system is better suited for situations of longer spans because product dimensions are not restricted by available stock sizes but are dictated by transportation issues and the erection method.

It is worth noting that the use of I-joists is not limited to floor construction. I-joists are also employed to form walls and roofs. Use in walls is generally an over-specification structurally. However, the increased depth of section facilitates increased levels of insulation and reduced cold bridging due to a thin web element that permits higher thermal performance to be achieved. In roof construction the use of I-joist results in longer clear spans and provides higher levels of insulation, resulting in a thermally efficient and sustainable roof structure. Off-site fabrication can also be employed depending upon the manufacturing, logistical and construction constrains.

Garage door opening

Examples of in-situ steel work

Figure 6.3 Examples of onerous load span conditions as a result of allowable depth of section

6.6 Combining timber with other materials

6.6.1 Timber and Steel

There are instances when large spans and heavy loads predominate while the available depth of the section is restricted [62]. In such instances, it is advantageous to specify a timber based product as it easier to connect to ancillary components and pre-install in the factory. The fitting of a steel section is awkward and time consuming site operation that also carries increased health and safety risks (*Figure 6.3*).

Serviceability criteria invariably dictate the specification, with deflection the governing criterion. The availability of solid timber sections is normally restricted to C24 at a maximum depth of 240mm and the thickness of the beam is restricted by the width of the bearing area. Therefore, specifying a larger timber section or higher strength grade is not normally an option.

There are other options available that avoid the need for a steel section. The use of a timber composite such as Intrallam (Laminated Strand Lumber or LSL), Parallam (Parallel Strand Lumber or PSL) or LVL are readily available with deeper sections. These products have a cost premium (being normally two to three times more expensive than C24 strength grade timber) and the mean E value that is used to determine stiffness EI in design is not generally an improvement on C24. Intrallam, Parallam and LVL have a mean E value of 10300, 12750 and 13500N/mm^2 respectively compared to 11000N/mm^2 for C24 timber. These products are therefore advantageous (given the additional cost) only if an increased depth of section can be specified.

Depth of section is normally restricted in design. If the beam is to form part of the floor system, it has to conform to the depth of joist being used. In housing construction the majority of cases are limited to a 241mm deep I-joist and, although there are occasions where deeper joists may be used for large spanning floors, the normal preference is to reduce joist spacing and maintain a shallow floor.

A further option is to use a flitch beam. This is a timber-steel-timber sandwich beam traditionally formed with a bolted connection (*Figure 6.4*). Flitch beams combine the benefits of timber construction (ease of working, readily available resource, simple connection of ancillary components) with the strength and stiffness of structural steelwork. Specification is governed by the stiffness of beam required and the allowable depth of section.

Flitch beams can also be fabricated using timber composites where additional stiffness is required and allowable depth is not limited. The most prevalent use of timber composites in domestic dwelling construction is as rim board material for floor cassettes. The specification of the rim board is at the discretion of the floor designer and supplier as they guarantee the flooring system. The use of composite timber materials as rim board material is mainly due to their stability as end products. The manufacturing process of timber composites results in a product with low moisture content (6-12%) which is strong and consistent, and less prone to shrinking, warping, cupping, bowing or splitting that occur in solid grade timber due to seasoning stresses.

Research on traditional bolted flitch beams dates back to 1859 [63]. The bolting together of the constituent parts is slow and inefficient and requires the pre-drilling of holes in the timber and steel for the elements to be bolted together. In 1973, Stern and Kumar [64] reported the use of hammer or machine-driven nails or gun-driven staples as an innovative method to connect the flitch beam elements together that would alleviate the problems associated with fabrication (*Figure 6.4*).

More recently, the use of hardened nails shot fired using a SPIT P200 cartridge gun has been investigated [65] [66] [67]. In order to ensure the structural robustness of flitch beams constructed from a more modern connection method, a range of qualified structural testing is required to prove the materials act in combination as required (*Figure 6.5*).

This alternative method of flitch beam fabrication employs readily available equipment that requires minimal training and maintenance and is more efficient due to speed of application. In many respects this reflects innovation through applied research and a lean approach. Through understanding employees, customer requirements, design and production procedures, an improved quality of product is achieved, which can be produced at a higher volume.

Off-site fabrication of a bolted LSL flitch beam

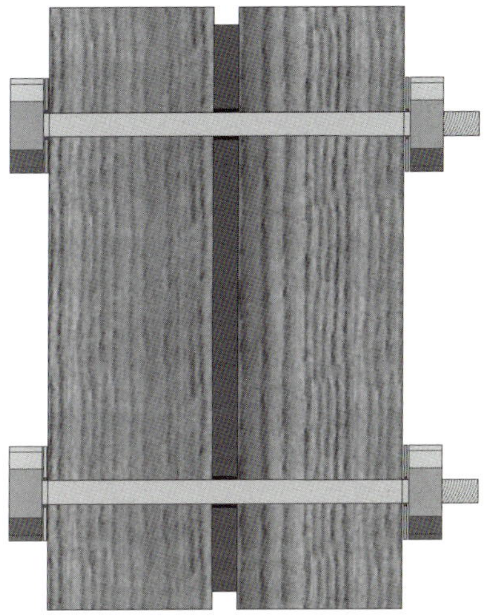

Bolted flitch beam detail

Nailed flitch beam detail

Figure 6.4 Bolted and shot-fired dowel flitch beams

Figure 6.5 Structural testing of nailed timber flitch beams

6.6.2 Timber and concrete

Timber-concrete composites are popular in some countries as a method of floor construction, particularly in multi-storey developments, as they provide a structurally efficient system which is both rigid and light. Timber-concrete composites are not limited to floor construction and can also be utilised for roofs and walls.

In floor construction, the timber members are utilised in the tensile zone with a concrete slab in the compression zone. Composite action can be achieved by a variety of connection systems (*Figure 6.6*) and, if the shear connection is efficient, beam depths can be significantly reduced when compared with a non-composite system. It is fundamental, however, that the long-term behaviour, particularly serviceability, is ensured as this depends on the creep of the concrete, the timber and the shear connection. Extensive research has been carried out in this area to prove the viability of timber concrete systems (68) (69) (70) (71). Also, in relation to serviceability, timber concrete composite floors more easily satisfy the vibrational performance requirements.

There is a developing interest in the use of timber-concrete composite construction in new buildings (*Figure 6.7*). This is motivated by the low embodied energy of timber, a need for thermal mass and the increasing cost of steel. Thermal mass is an important attribute in terms of passive heating and cooling. It uses the heat capacity of the building itself to moderate internal temperatures, and therefore to reduce the energy required for heating and air-conditioning. Concrete has a high capacity to retain heat and consequently concrete buildings have an inherently high thermal mass. Timber-concrete composite construction allows structurally-efficient use of both timber and concrete whilst retaining the thermal mass of a concrete floor slab.

Timber-concrete composite flooring systems can provide a higher degree of sound insulation than solid timber systems due to their greater mass and,

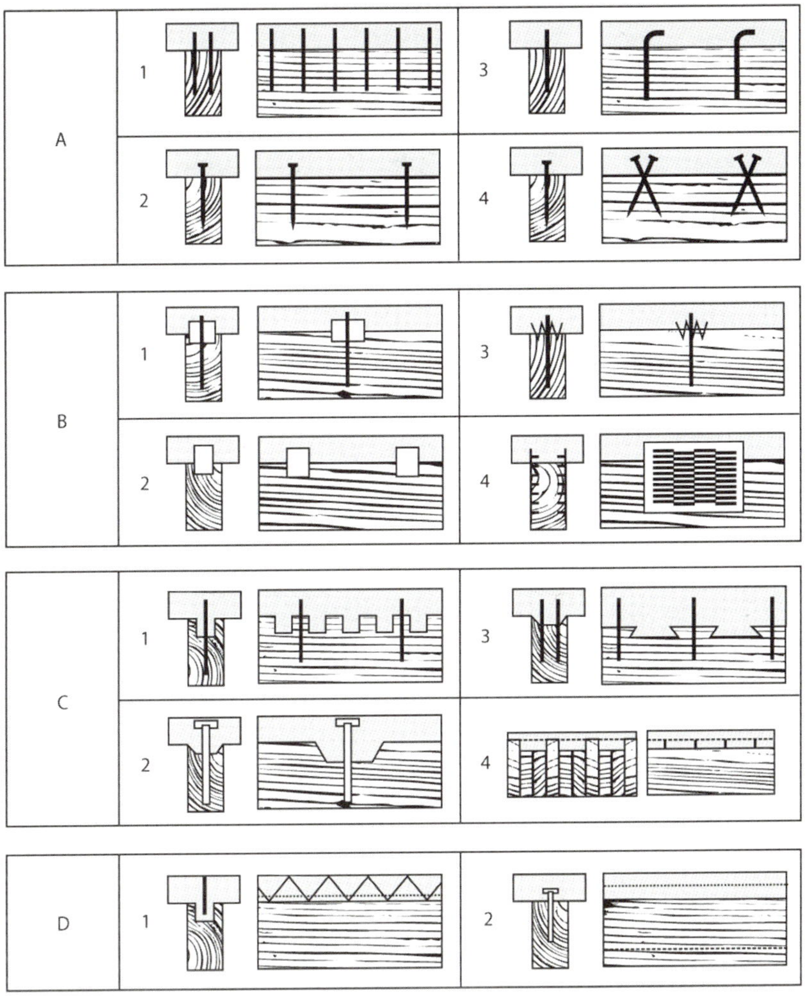

(A) nails, screws or dowel type fasteners
(B) surface connectors
(C) notched connectors
(D) bonded connectors

Figure 6.6 Examples of different types of timber-concrete connections [91]
After Ceccotti

as a result of higher damping, insulation to impact noises is also improved. The upper concrete slab also acts as an efficient barrier to fire propagation and, if connection detailing is specified appropriately, can improve overall structural performance by optimising the use of materials. This because,

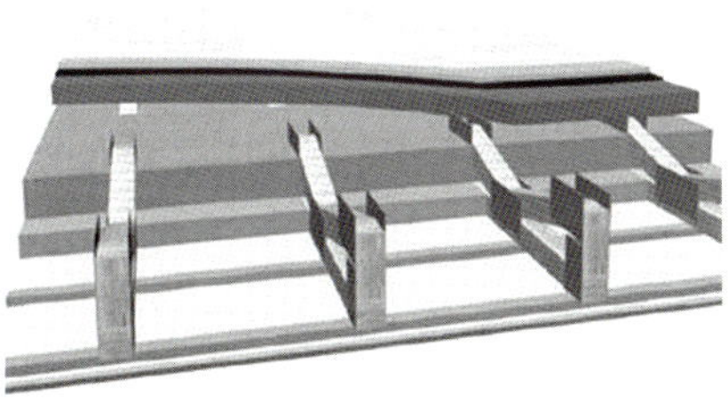

Wood-concrete floor system

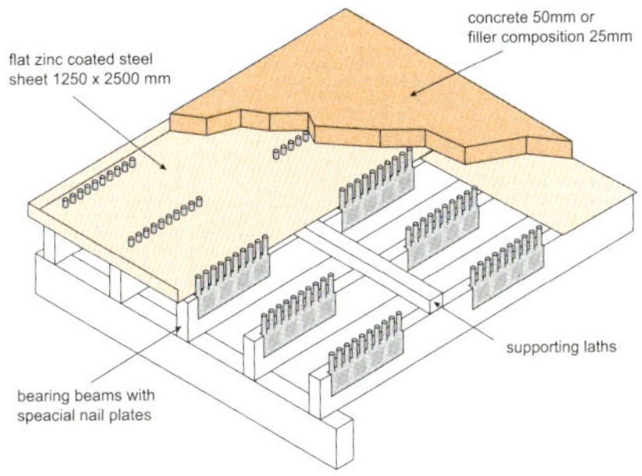

Light structured slab

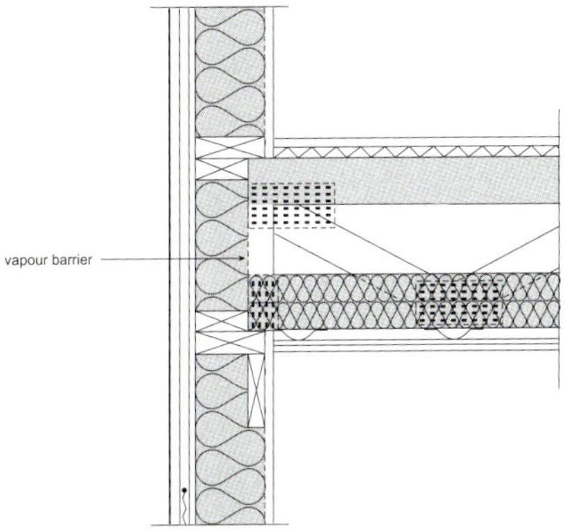

An example of a floor - supporting wall connection showing the vapour barrier position.

Figure 6.7 Timber concrete composite floor system using nail plates for shear connectors
Drawings: SEPA Oy

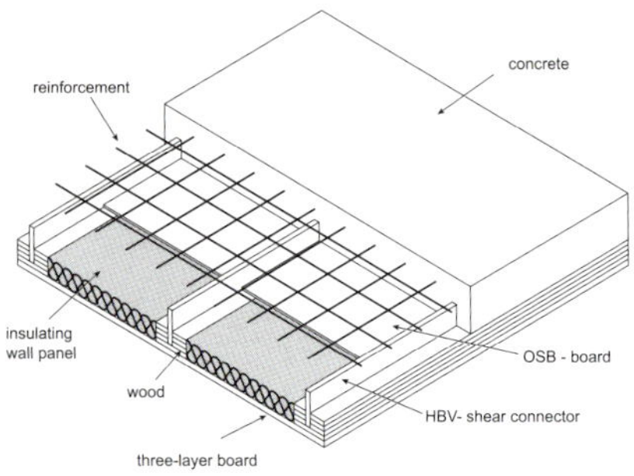

Figure 6.8 Detailing and fabrication of timber concrete composite 'Box' floor system
Drawing: TiComTec GmbH

in bending, the concrete screed of the top chord carries the compression stresses, with the timber section in the bottom chord providing resistance to tension stresses.

A range of drawbacks exist in situations where wet concrete is cast on top of timber beams in-situ with mounted connections. These include the time needed for the concrete to cure, lower stiffness and higher creep during concrete curing, the higher cost of cast-in-situ concrete, and some possible problems of quality control. Further to this, as concrete cures it shrinks and this induces tension on the timber at the bottom cord and results in deflection.

Off-site prefabrication offers one solution to these drawbacks (*Figure 6.8*). A notable benefit of prefabricating the concrete slab is that most of the concrete shrinkage will develop when the slab is not yet connected to the timber beam, markedly reducing the correlated increase in deflection and flexural stresses in the composite beam [72]. Another advantage of the prefabrication is the reduction in construction cost by moving work from the building site to the workshop. This improves the efficiency and quality of the construction process as well as presenting opportunities to save resources and simplify the recycling of waste.

From a financial perspective, timber-concrete composites are competitive when compared to all concrete systems. A holistic cost-benefit approach measures the overall benefits in terms of improved execution time, reduced formwork and stabilisation requirement (the timber elements can provide this during construction). Reduced foundation costs as a result of a more lightweight system also have to be considered. The viability of timber-concrete composites depends upon the nature of the project.

6.6.3 Structural insulated panels (SIPs)

Reconstituted timber in combination with insulation (structural insulated panels or SIPs) has been used for both improved building and structural performance.

SIPs are a sandwich construction of OSB and expanded polystyrene (EPS) that provides both structural support and insulation in a single system. They are normally used for wall and roof construction. The concept of SIPs began in 1935 at the Forest Products Laboratory (FPL) in Madison, Wisconsin. FPL engineers speculated that plywood and hardboard sheathing could take a portion of the structural load in wall applications.

The architect Frank Lloyd Wright used structural insulated panels in some of the affordable Usonian houses he designed in the 1930s and 1940s. SIPs took a major leap in technology when one of Wright's students, Alden B. Dow, son of the founder of Dow Chemical Company, created the first foam core SIP in 1952. By the 1960s, rigid foam insulating products had become readily available and resulted in the production of structural insulated panels as they are known today.

Unlike stress skin panels that can be made by on-site skilled tradespersons, SIPs are manufactured off-site and require quality-manufacturing standards to ensure that the product can perform as intended.

SIPs panels in a press

SIPs wall connection detail

Figure 6.9 Structural Insulated Panels

There are two typical panel fabrication techniques:

- Adhesive is applied to a pre-cut foam core, then the core is cold pressed between two facing boards (*Figure 6.9*).
- Foam is poured into pre-spaced facings and the foam cures to bond the facings, this procedure is referred to as 'autohesively bonded'.

SIPs are also typically lightweight, although this is dependent on the outer-skins. The light weight nature of SIPs facilitates on-site installation and, as a result of the insulation being preinstalled off-site, SIPs tend not to suffer from sagging or wet insulation (due to exposure on site), which can reduce thermal performance. Nor are SIPs prone to gaps and voids in insulation coverage left by poor site workmanship. *Figure 6.10* shows a sequence for installation of a SIPs building.

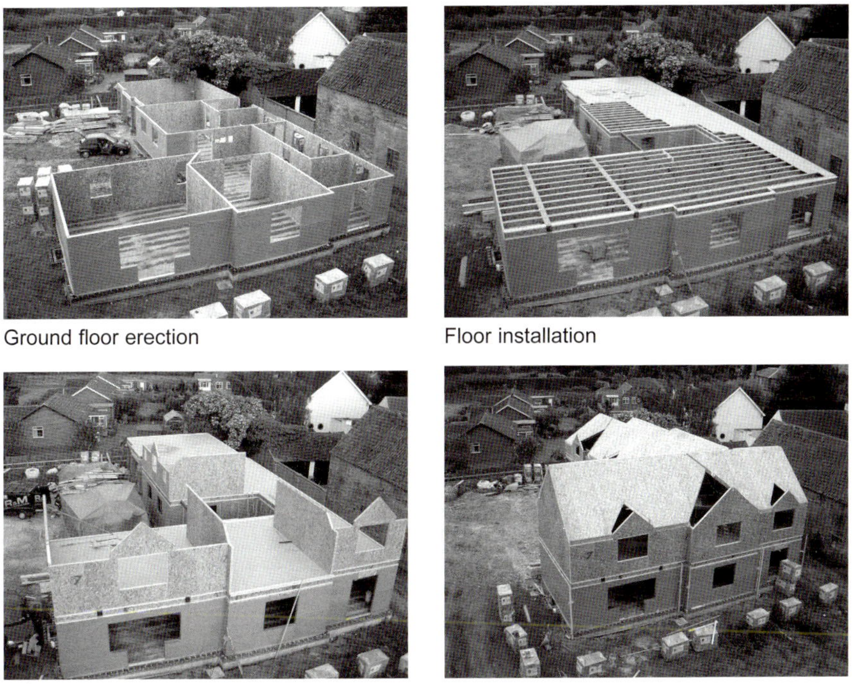

Ground floor erection

Floor installation

1st floor erection

Roof installation

Figure 6.10 Construction of a SIP system in the UK

From an environmental perspective, oil based insulating foams can be melted down and reused. The insulations are also very low density and consequently only small masses of the primary resource are needed to produce high levels of insulation. OSB is produced from fast-growing trees, forest thinning and recycled timber. In conjunction with this, when OSB boards are autohesively bonded to the rigid urethane insulation core during manufacturing, the requirement for potentially environmentally harmful adhesives is eliminated.

The environmental credentials of SIPs, however, are not so much in the constituent materials but in the potential energy efficiency of the building formed from them. A typical external wall construction comprising SIPs with cores of autohesively bonded polyurethane, finished with 12.5 mm plasterboard on timber battens and a brick outer leaf with a 50 mm vented cavity, can achieve an estimated U-value of between 0.19 and 0.22 W/m^{2}K depending on the number and type of panels.

SIPs also have a lower level of thermal bridging as a result of the reduced requirement for structural elements such as wall studs and lintels. For a SIP to function structurally, it relies on the composite action between the skins and the core, a requirement of which is a robust bonding method. Extensive research over the last few years has shown that, when manufactured under appropriate conditions, a suitably strong bond can be achieved. There has also been extensive research qualifying the performance of SIPs in terms of vertical loading (direct compression), transverse wind loading (combined bending and compression), in-plane lateral forces (racking loads) and the effects of medium-term loading on the deformational characteristics (creep effects)[73] [74].

In terms of longevity and continued structural performance, there are SIP buildings in the USA that have been in service for 50 years. Therefore, the structural integrity of SIPs has been demonstrated both through testing and service, providing accredited manufacturing and erection methods are adhered to. There is, however, limited available information on buildings over three storeys that use SIPs as the structural load bearing walls and, as a result, their application is limited somewhat to low rise construction.

Fire regulations must be considered when using SIPs. To achieve current fire regulations for residential dwellings for the UK, the internal linings of the structure require a class 0 (non-combustible) or class 1 (semi-combustible) lining depending on the size and occupancy of the building relative to the required fire protection. This can be achieved by providing one layer of 12.5mm gypsum plasterboard to obtain class 1 and two layers to obtain a class 0 fire rating.

The added value of SIPs does tend to come at a cost, with SIPs generally costing 2-10% more than an insulated and sheathed wood frame. However, with wood framing, additional insulation requires deeper lumber dimensions or double framing. SIP insulation, in contrast, gets less expensive per unit volume as the panels get thicker, since the skins and manufacturing and installation process remain the same. It should also be noted that off-site application of insulation can result in reduced waste and an improved quality of product. Therefore, when high levels of insulation are required, SIPs are at an advantage to other forms of construction although, in terms of meeting current building regulation requirements, they tend to be at a cost disadvantage.

SIPs are therefore a legitimate alternative when considering low rise construction and are also an ideal modern method of construction. SIPs have been proven to exhibit a level of structural robustness that meets and exceeds building regulation requirements and, when considering the building envelope, they have improved building performance properties. SIPs can also be combined with other timber engineered products, as demonstrated by the Structural Insulated Roofing System (SIRS) developed by Donaldson Timber Engineering (*Figure 6.11*). SIRs is a room-in-roof system that combines floor cassettes, double ply support trusses and structural insulated panels. The system can be used to form the majority of roofing shapes including; gable-to-gable, Dutch hip, T-shaped and dog-leg but excluding hipped roofs.

External walls erected ready for floor SIPs

Installation of floor cassette

Installation of spandrel panels

Erection of roof trusses

Fitting roof SIP

Structural Insulated Roofing System (SIRS) complete with dormers

Figure 6.11 Erection of a SIP system house (Structural Insulated Roofing System or SIRS)
Photos: Donaldson Timber Engineering

6.6.4 Timber and carbon fibre

More recently, improving the structural performance of timber by means of combining it with another material has been carried out using carbon fibre reinforced polymers. Gilfillan et al (75) write that such a process can significantly enhance the performance of low grade timber in the form of Carbon Fibre Reinforced Wood (CFRWood). However, the financial viability of such a process is still in question and the use of such technologies is restricted to bespoke projects and the retro-fitting of degraded timber elements in dilapidated structures.

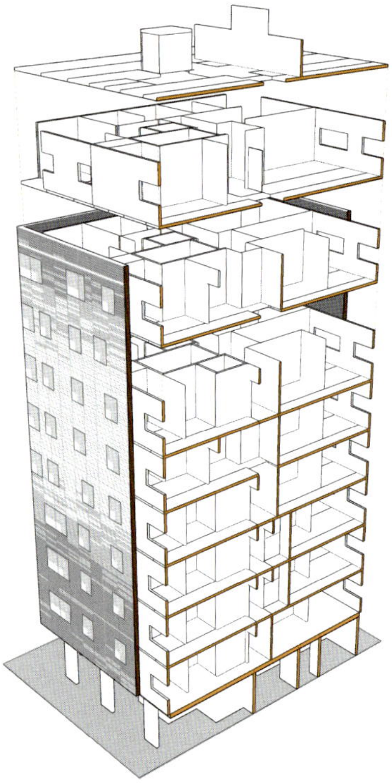

Cellular structure

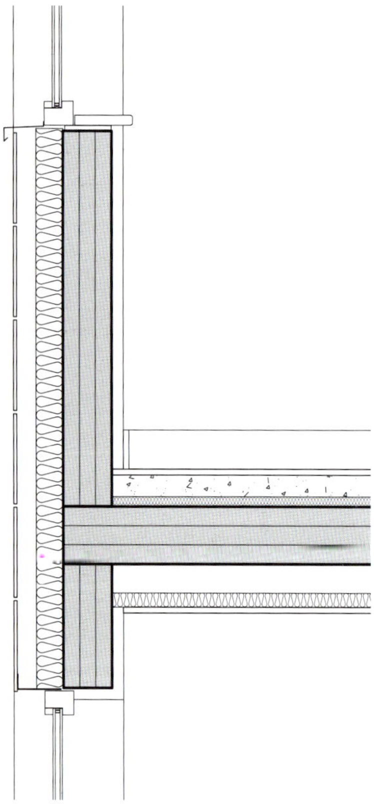

Construction detail

Figure 6.12 Nine-storey residential tower from cross-laminated solid timber panels at Murray Grove, London
Drawings: Waugh Thistleton Architects

6.7 Case Study: Re-engineered multi-storey massive timber design solution

The Stadthaus apartment building constructed from KLH manufactured cross laminated solid timber panels in Murray Grove London is an example of the use of re-engineered timber providing a timber building for 9 storey accommodation. In this project cross laminated timber panels form a cellular structure of timber load bearing walls, including all stair and lift cores, with timber floor slabs (58) (*Figure 6.12*).

In order to achieve the nine-storey construction a range of innovative techniques were used to resolve pre-conceived limitations. Because panellised systems are more susceptible to progressive collapse, Techniker, the structural engineer, carried out a rigorous assessment of design options to ensure the robustness of such a tall timber structure. This research into the movement characteristics of tall timber forms is continuing. In the case of the Stadthaus project, suitable redundancy was achieved by means of strengthening the ties between units to a sufficient level to resist anticipated accidents, blast loads or unexpected impacts.

Techniker's preferred option has been to exploit the over-structuring typical in residential layouts by conjecturing alternative load-paths should any component be compromised. As a result a policy of 'efficient redundancy' was employed and, wherever possible, floor panels are designed to span in two directions or to cantilever if a support is removed. Effective ties are provided between floors and walls using simple 'off-the-shelf' brackets and screws. The inherently high in-plane stiffness of the cross-laminating process provides 'built-in' redundancy in the form of wall elements which can span laterally if support beneath is removed.

With the system being nine storeys, other structural timber engineering challenges had to be met including overcoming the 'weak link' of the wall, floor and wall junction. The end grain strength of the wall panels is 24 N/mm^2 while the cross-grain crushing capacity of the floor plate is only 2.7 N/mm^2. As a result screw and nail arrays were employed at points of high load intensity to alleviate stress concentrations (76) (*Figure 6.13*).

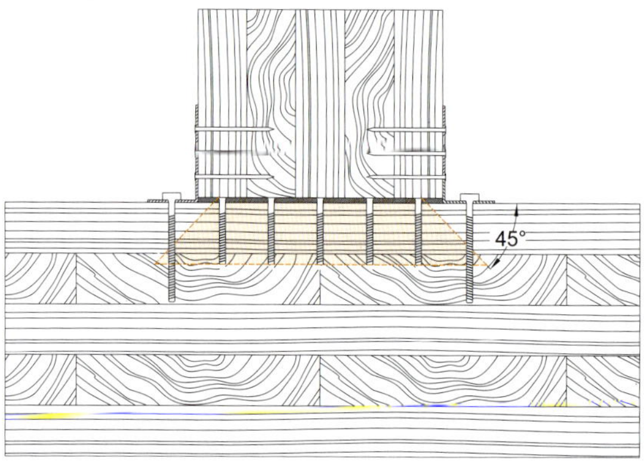

Figure 6.13 Floorplate nailed to alleviate local crushing

7 Summary

The modern construction industry is faced with major challenges, the primary focus being how to achieve more sustainable forms. In order to be sustainable, more is required for less; therefore efficiency, innovation and optimisation are necessary. Modern Methods of Construction are about delivering these objectives.

MMC do not necessarily have to be carried out off-site, but key principles need to be adopted in order to:

- make the process more transparent
- engage the work force to nourish a culture of innovation and continuous improvement.

Innovation can address customer needs not just by supplying an optimised product, but also a service that improves value.

Moving more of the processes to an off-site controlled environment allows the theories of lean production to be applied. By implementing a production process, the non-value adding activities can be more easily identified and, through applying a holistic management philosophy, more readily rationalised. On site, the concept of lean is more difficult to implement because the project-focused framework is not necessarily organised in a flowing manner. Nevertheless, there are methods of achieving a more lean approach such as increased visualisation, daily huddle meetings and the 5S Process.

Construction projects will always be a combination of off-site and on-site activities, the relative levels of which will depend on the project itself and systems being used.

A key MMC concept is a constant striving for perfection whilst managing the balance between standardisation and flexibility to ensure project success. In order to achieve this there are many timber-based products and system solutions available, all evolving in response to culture, demand and available resource. What is specified will depend on the project requirements and procurement strategy. In housing terms: is the building a one-off project for a self builder or is it to be supplied repeatedly to a volume house builder?

Adopting lean principles will improve the sustainability performance by making better use of resources and reducing waste. This said, construction should not be considered as just a linear process to achieve an end goal of project completion. It is actually a cyclical process that can become more sustainable by taking account of building use, re-use and re-cycling.

Energy is required for felling timber, processing it and conversion to the end use. Therefore the embodied energy of the full chain of custody needs to be considered. Chain of custody schemes ensure timber is being sourced responsibly.

Energy is also required during the construction process and, once constructed, the building is required to minimise energy use. More demanding building regulations set minimum standards for conserving energy. There are different ways of achieving this such as enhancing the thermal performance of the building fabric or generating energy by means of adopting renewable methods. Timber MMC have demonstrated ability to deliver the higher thermal standards demanded by the Code for Sustainable Homes and also the capability to perform to Passive House requirements.

It is not only environmental legislation which is changing in favour of the adoption of timber MMC. The Eurocode for structural timber design is also driving change and will promote innovation. The revised code of practice will facilitate the specification of non-standard timber systems as a result of being more transparent and adaptable. And the engineer will be able to rationalise standard construction details to reduce cold bridging and air leakage, thereby improving building envelope performance.

Table 7.1 summarises the advantages and disadvantages of four generic timber building systems that can be employed in an off-site MMC strategy.

Modern methods of timber construction are therefore not only about new products. Off-site and modern methods of timber construction are about improving the overall process from tree growth, production, design, construction, use and re-use.

Improving the whole process requires decision making to be transparent and interlinked in order that improvements can be continuous and optimal. In order to do this, software-based systems such as Enterprise Resource

Table 7.1 Generic advantages and disadvantages of off-site MMC solutions

System	Advantages	Disadvantages
Closed panel timber frame solution	• High levels of prefabrication can be carried out off-site including the installation of insulation, services, external cladding and internal lining. • Lightweight systems facilitate the erection process and also reduce foundation requirements. • With good detailing a high level of air tightness can be achieved. • Fire protection is provided by the application of fire retardant sheathing materials such as plasterboard. The pre-installation of services can also include a sprinkler system for active protection once built.	• Limited in terms of multi-storey construction due to design limitations (settlement) and robustness (panelised buildings are susceptible to progressive collapse due to a lack of redundancy). • High levels of mechanisation and the use of standardised components can limit flexibility. • Limited thermal mass. Due consideration should be taken in design especially when specifying highly airtight systems with solar gain, as overheating can be a problem. • Tolerances are critical especially in highly serviced systems.
Massive timber systems	• Excellent structural credentials being inherently more rigid. • Cross lamination results in a two-way spanning horizontal slab. • Additional fire protection as a result of charring. • Less susceptible to movement. • Panels can be formed and cut in a factory to the required shape and size. • Routing and holes for services and connections can be incorporated within the panels.	• Over specification of material for small scale low rise systems. • Requires a large amount of raw material and thick wall sections in order to achieve building performance requirements. • As a panelised system, due consideration is required in design to ensure system robustness. • High levels of service inclusion in the factory may require increased levels of capital investment in CAD/CAM and CNC equipment.
Beam and post system	• A high degree of flexibility can be achieved whilst only needing a small range of standard elements and connection details. • Additional fire protection can be achieved through charring.	• Can require a higher degree of on-site construction. • Connectivity between elements is critical to ensure system rigidity and robustness. • Inclusion of services in the factory is limited unless non-load bearing panelised systems are fabricated for inclusion.
Hybrid systems	• Optimises use of component parts and connections in order to gain higher levels of performance.	• Systems can be more complicated and, as a result, this can impinge upon levels of flexibility. • Hybrid solutions are often new and 'innovative' and, as a result, can have a higher degree of associated risk.

Planning will become more necessary. ERP will combine the processes and enable continuous monitoring so that waste can be identified and reduced, workers can be more engaged in the processes and there is improved communication between suppliers and clients.

Timber is well placed to meet the challenges of a modern, sustainable construction industry through the adoption of MMC.

MMC fundamentally result in suppliers becoming vendors, workers being team players and clients becoming partners within a continuous improvement cycle, where applied research and development drive an evolutionary process.

Appendix A: The Code for Sustainable Homes

The Code for Sustainable Homes (CSH) is a design guide that was produced with the aim of helping UK housing developments to achieve zero carbon emission levels by 2016. CSH considers environmental performance as a two stage process (design stage and post construction). It identifies nine key impact categories and associated issues that impact on the environment. From these, key performance measures have been set (*Table A.1*) to reduce environmental impacts and exceed legislation and regulatory requirements.

The Code uses a sustainability rating system – indicated by 'stars' – to communicate the overall sustainability performance of a home. A home can achieve a sustainability rating from one (♦) to six (♦♦♦♦♦♦) stars depending on the extent to which it has achieved Code standards. One star (♦) is the entry level – above the level of the Building Regulations; and six stars (♦♦♦♦♦♦) is the highest level – reflecting exemplar development in sustainability terms.

The sustainability rating that a home achieves represents its overall performance across the nine Code design categories. Minimum standards exist for some categories (*Table A.1*) and these must be achieved to gain a one star (♦) sustainability rating. Energy efficiency and water efficiency categories also have minimum standards that must be achieved at every level of the Code in recognition of their importance to the sustainability of any home. Provided the minimum requirements are met for certain categories, developers can then choose which and how many standards they implement to obtain 'credits' under the Code in order to achieve a higher sustainability rating.

CO_2 emissions

Taking the example of energy/CO_2 emission rates, the emissions rating for a dwelling (known as the Standard Assessment Procedure (SAP) rating) will be assigned a level (*Table A.2*). The SAP rating target can be met in a variety of ways – eg improving the fabric of the building, improving the efficiency of lighting and space heating, and by using lower carbon fuels – while the Code also covers the energy/CO_2 emissions of a small range of white goods (freezers, dishwashers, etc). [77].

The Code Level to achieve a one star (♦) rating is a 10% improvement on the requirements enforced following the 2006 Building Regulations for England and Wales, the aim of which was to reduce CO_2 emissions from new homes by 20% compared with the previous (2002) standards. (Scotland and Northern Ireland have different Regulations as well as different time scales for implementation.) The purpose of the Building Regulation requirements is for all new buildings in England and Wales to meet minimum mandatory standards (the Target Emission Rate).

Green Guide to Specification

The CODE assessment process also incorporates The Green Guide to Specification (previously incorporated into BREEAM and EcoHomes). The Green Guide to Specification is used to consider the embodied environmental impacts of different specification options by evaluating their impact

Table A.1 Summary of environmental impact categories and issues

Categories	Issues	Code Flexibility
Energy and CO_2 emissions	Dwelling emission rate (M) Building fabric Internal lighting Drying space Energy labelled white goods External lighting Low or Zero Carbon (LZC) technologies Cycle storage Home office	Minimum standards at each level of the Code
Water	Internal water use (M) External water use	
Materials	Environmental impact of materials (M) Responsible sourcing of materials – building elements Responsible sourcing of materials – finishing elements	Minimum standard at Code entry level
Surface Water run-off	Management of surface water run-off from developments (M) Flood risk	
Waste	Storage of non-recyclable waste and recyclable household waste (M) Construction waste management (M) Composting	
Pollution	Global Warming Potential (GWP) of insulants NOx emissions	No minimum standards
Health and Wellbeing,	Daylighting Sound insulation Private space Lifetime homes (M)	
Management	Home user guide Considerate constructors scheme Construction site impacts Security	
Ecology	Ecological value of site Ecological enhancement Protection of ecological features Change in ecological value of site Building footprint	
(M) Denotes issues with mandatory elements.		

on the Life Cycle Analysis of the whole system. It awards credits for using specifications that minimise environmental impact. The 'big picture' of this approach is shown in *Figure A.1*.

Table A.2 Code levels for Mandatory Minimum standards in CO_2 emissions

Code Level	Minimum Percentage reduction in Dwelling Emission Rate Over Target Emission Rate
1 (♦)	10
2 (♦♦)	18
3 (♦♦♦)	25
4 (♦♦♦♦)	44
5 (♦♦♦♦♦)	100
6 (♦♦♦♦♦♦)	'Zero Carbon'

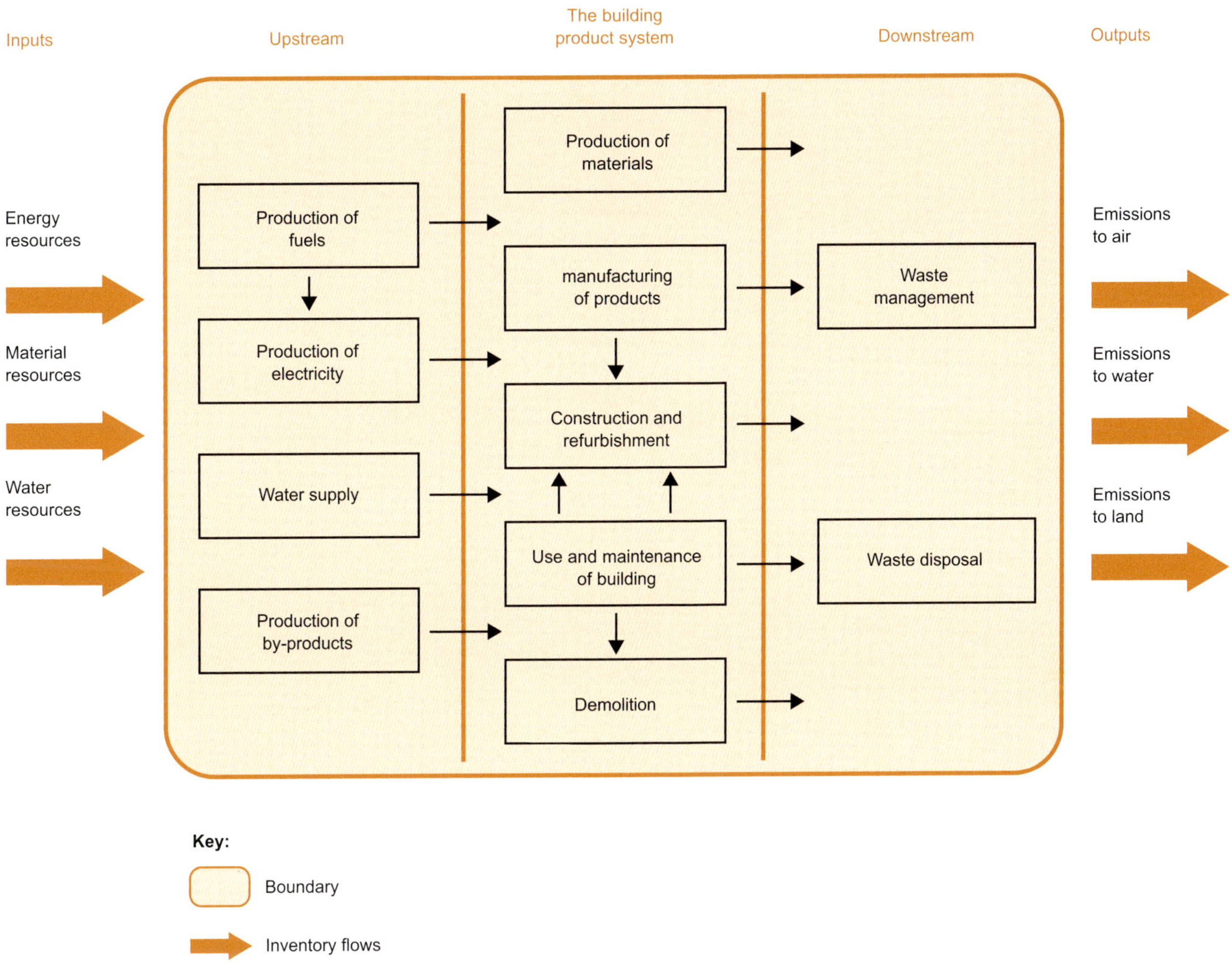

Figure A.1 Life cycle system of a building and its interventions with the physical environment [93]
After Anderson et al

Assessment

The assessment process (*Figure A.2*) checks firstly that the four mandatory standards are met. Then the 'tradeable' credits and weighting factors are added, giving a total percentage score, which designates the Code Level. The Code assessment is recorded on a certificate assigned to the dwelling.

Renewable energy

Code credits can be gained by incorporating renewable energy technologies (*Table A.3*). In order to achieve Code Levels 5 and 6, it is necessary to generate electricity either communally or through micro-generation on each individual house. Construction cost reductions are more attainable on higher density developments as there is greater potential to employ site-wide carbon saving technologies. To achieve Code Level 6, microgeneration technologies are generally required (eg photovoltaics). Although this adds to the overall cost, it is predicted that costs of compliance will fall over time and that by 2016 they could be reduced by 16-25% depending on Code level [78]. However, the cost of microrenewables remains an issue where the returns on investments are yet to be fully established and, in some cases, exceed the working life of the technology. Practical considerations, such as

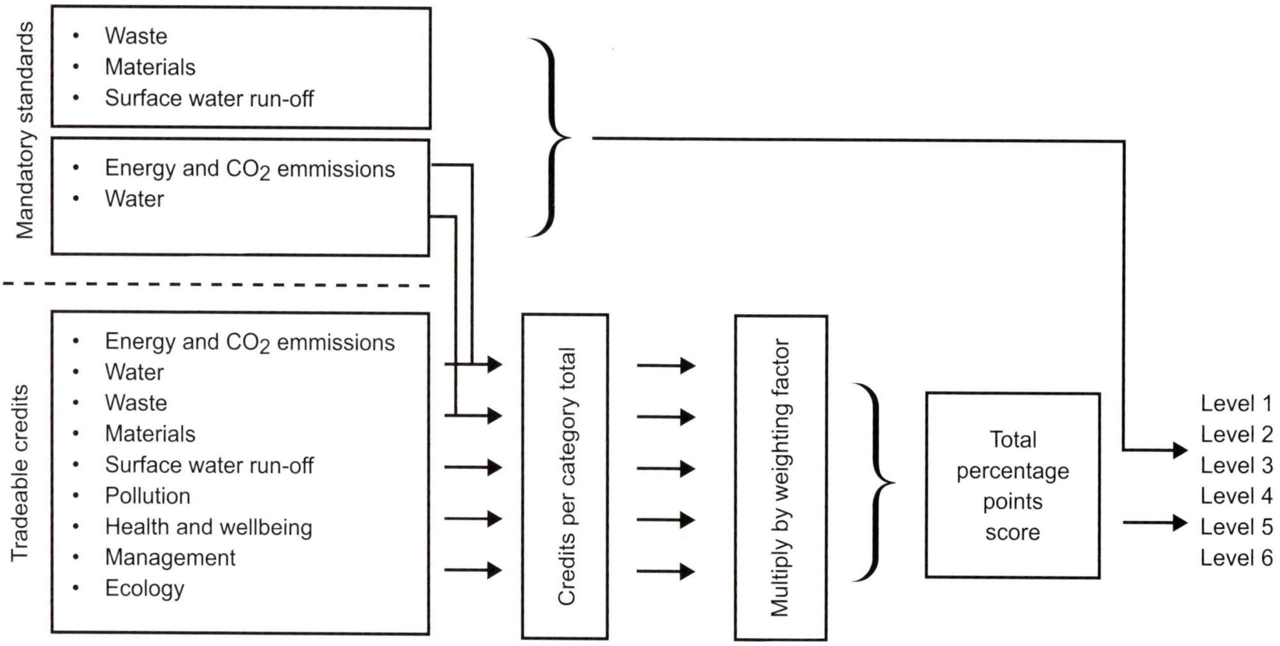

Figure A.2 Scoring system for the Code for Sustainable Homes (96)
After DCLG

Table A.3 Renewable Technologies (79)

Micro-renewable	Overview
Solar thermal (solar hot water)	Solar thermal or solar hot water (SHW) systems for use in dwellings use a heat collector which is generally mounted on a roof, and contains a fluid (usually water with glycol) which is heated by the sun. The heated liquid is then passed through a coil in a hot water storage cylinder. The water in the cylinder may then be supplied directly, or raised to a higher temperature (if required) by a boiler or electric immersion heater
Solar electric	Photovoltaic (PV) systems convert energy from the sun into electricity through semi-conductor cells. A cell consists of a junction between two thin layers of dissimilar materials, usually silicon based. When light shines on the junction, a difference in energy is created – otherwise known as 'voltage' or 'potential difference'. This voltage is used to produce an electrical current or direct current (DC), which can be used directly or converted into alternating current (AC) depending upon the application
Ground source heat pumps	A heat pump extracts heat from the ground, air or water and transfers it to a heating distribution system, such as underfloor heating, using an electric pump. Ground source heat pumps (GSHP) are currently the most common type of heat pump used for domestic space heating in the UK, and use technology which is essentially the same as that in a domestic refrigerator. A typical GSHP system will comprise a ground heat exchanger (for extracting heat from the ground), the heat pump itself and a heat distribution system.
Biomass	Biomass is a term used to define all plant and animal material and has been used as an energy source for centuries. Although there are a number of different technologies available that extract heat from biomass, wood burning systems are most likely to be appropriate for use in individual dwellings. Wood is an extremely versatile energy source and can be burned in a number of different forms and appliances. Biomass or wood burning systems differ from other renewable energy sources in that CO_2 is emitted when it is burned to produce heat. However, the amount of CO_2 released is only the same as the amount of carbon absorbed by the tree whilst it was growing. Using biomass as a fuel is essentially a carbon neutral approach. But the fuel can only be called a true renewable energy source if the fuel has come from a sustainable source (ie it is processed and replenished). It should also be used close to where it was originally grown to reduce secondary CO_2 emissions incurred through transportation
Wind	The UK has the largest potential wind resource in northern Europe with approximately 40 per cent of the total supply available. This potential is recognised by the UK Government which sees wind energy as making a significant contribution to its renewable energy target of supplying 10 per cent of total energy requirement by 2010. Wind turbines can have outputs ranging from a few watts to several megawatts and produce electricity without emitting CO_2. Energy is extracted from the wind using a rotor generally consisting of two or three blades which have a profile similar to that of an aeroplane wing. If the diameter of the rotor is doubled, the power output from the turbine is quadrupled at a given wind speed.

installation and maintenance of all renewable energy technologies, are yet to be fully addressed, while local site topography can have a large bearing on the efficiency of wind power generation.

Water consumption

CSH home compliance also has minimum mandatory water use requirements of 105 litres per person per day for Code Level 3 and 4 and 80 litres per day for Code Levels 5 and 6. Possible methods of achieving the water consumption requirements include low-water use items such as dual-flush toilets, flow-regulated or flow-restricted taps and showers and low-use white goods with harvested or recycled water.

Appendix B: Timber design properties

Density

Density is considered the most important physical characteristic of timber. Most mechanical properties of timber are positively correlated to density as is the load carrying capacity of timber joints, [80].

At a macro level, slow growth rate and consequent closeness of grain results in thick-walled tracheids which, being tightly packed, result in higher strength. The cell structure of softwoods consists of 90-95% tracheids, [80], which are normally 25-45μm in diameter and 3–5mm in length. The tracheids are aligned radially, which results in softwood being strongly anisotropic (strength dominant in the direction of the grain) [81]. This anisotropic nature is enhanced by the fact that cell wall thicknesses vary as do the size of cavities. The larger the cavity, the better the cell is at conduction of water and, conversely, the thicker the walls and the smaller the cavity the less suitable the cell is for conduction but the better it is at providing strength [82].

A solid piece of wood, ie a piece only consisting of the cell wall material, in which there are no cell wall cavities and intercellular spaces would have a density of 1500kg/m^3. As a result of the different cell wall to air space ratio of different species of timber, the density of wood ranges from 160kg/m^3 to 1250kg/m^3, [82]. Timber porosity therefore determines the density and also gives an estimate of the amount of water that can be held. *Figure B.1* illustrates the cellular and structural features of wood.

Moisture content

Wood is hygroscopic, that is it continually exchanges moisture with the surrounding atmosphere. It therefore attains a moisture content in equilibrium with the water vapour conditions of the surrounding atmosphere. This condition is called the equilibrium moisture content. Wood is also mechanosorbic, meaning its structural properties are influenced by the amount of moisture it holds. An increase in moisture content produces lower strength and elasticity values.

Fibre saturation point is the point at which only bound water remains, as the free water has been removed. Bound water is held within cell walls by hydrogen bonds and van der Waals forces. Free water exists in the cell lumen and cavities in liquid and/or vapour form. The removal of bound water requires much more energy than the removal of free water.

When moisture is removed from the cell wall, timber shrinks. Shrinkage and swelling within the normal moisture range for timber structures are termed 'movements'. The problems of dimensional movement of timber are therefore resolved by using timber of a moisture content corresponding to the relative humidity of its environment.

Most mechanical properties of defect-free wood improve with decreasing moisture content below the fibre saturation point. It is generally accepted that strength increases overall with reduction in moisture because of the

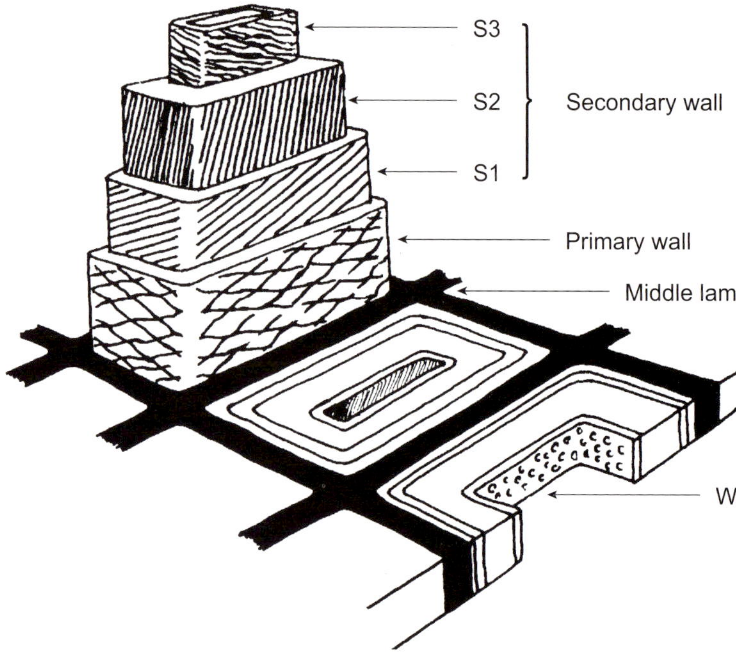

Cell wall organisation of a mature tracheid (90)
Diagram: Dr M D Hale

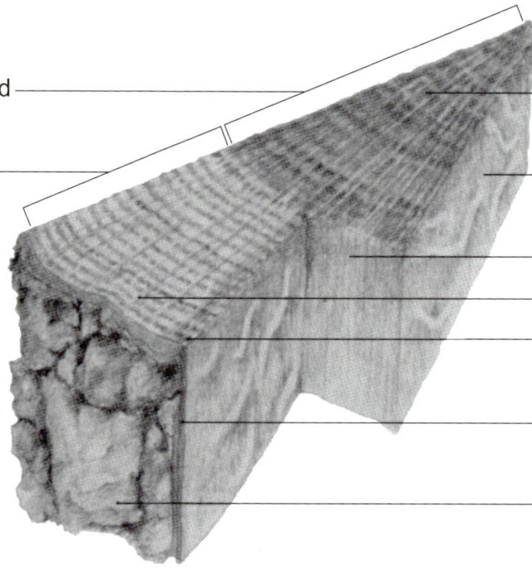

Diagrammatic representation of a wedge shaped segment cut from an oak tree showing the principal structural features (83)
Diagram: TRADA Technology

Figure B.1 Cellular and structural features of wood

shortening and consequent strengthening of the hydrogen bonds that link together the microfibrils that make up the cell wall in three layers (S1, S2 and S3) as shown in Figure B.1a.

The overall relationship between mechanical properties and moisture content is approximately for wood in the range 8-22% moisture content.[81] *Table B.1* shows the average changes in mechanical properties of clear wood due to a 1% change in moisture content. Moisture variations above the fibre saturation point have no effect on mechanical properties since such variations are related to free water.

Table B.1 Approximate increase (%) of clear wood properties for a one percentage fall of moisture content [80]

Property	Increase (%)
Compressive strength parallel	5
Compressive strength perpendicular	5.5
Shear strength parallel	3
Modulus of rupture parallel	4
Modulus of elasticity parallel	2
Tension strength parallel	2.5
Tension strength perpendicular	1.5
Note: Properties at 12% moisture content form the datum.	

Temperature

At temperatures within the range +200°C to -200°C, and at constant moisture content, strength properties are linearly (or almost linearly) related to temperature, decreasing with increasing temperature [83].

Time

Timber is a viscoelastic and rheological material. Viscoelasticity means that the behaviour of the material is time-dependent; at any instant in time under load its performance will be a function of its past history. Rheology indicates that its behaviour is also a function of its thermal and moisture histories and their interaction with the loading history.

The modulus of rupture (maximum bending strength) will decrease in proportion, or nearly in proportion, to the logarithm of the time over which the load is applied; failure in this particular time-dependent mode is termed creep rupture or static fatigue [83].

Like some other materials, wood exhibits three creep phases under constant stress:

1. Primary creep is an initial phase during which the rate of deformation accumulation decreases with any increase in elapsed time.
2. Secondary creep is a phase during which the rate of deformation accumulation is constant.
3. Tertiary creep is a phase during which the rate of deformation accumulation increases with any increase in elapsed time.

Whether secondary or tertiary creep phases are entered depends upon the stress level and the elapsed time. At high enough stress and temperature levels, all three stages will occur and result in fracture of the material at the end of the tertiary stage [84].

Grain deviation

Grain deviation is an important influencing factor on timber strength properties. In straight-grained timber the fibres or tracheids are more-or-less parallel to the vertical axis of the tree. This results in a high strength of timber in tension along the grain, with a low strength perpendicular to the grain, approximately 48:1.

Knots

There are two different types of knot to consider: tight and loose. Tight knots are intergrown with the surrounding tree as a result of the girth of the trunk increasing and successive growth rings forming over the stem and branches. If the limb dies or is broken off, subsequent growth rings added to the main stem simply surround the dead limb stub and the dead part of the stub becomes an encased knot. Because it is not intergrown and often has bark entrapped it is called a loose knot.

Knots have a negative effect on most mechanical properties of wood because they distort the flow of the grain. Consequently eccentricities inevitably develop in the flow of forces within components containing knots. Whatever the nominal stress condition for a timber component, there will be stress components perpendicular to grain and shear stress parallel to grain. How critical this is depends on the positioning of the knot(s) within a component as well as its size, soundness, and geometry.

Strength class and grading

Strength classes are used to assist in the process of design. A strength class system groups together grades and species with similar strength properties thus making them interchangeable. This then permits an engineer to specify a chosen strength class and the characteristic strength values of that class in design calculations.

Traditionally timber grading was carried out by visual inspection and the most important strength determining factors were rate of growth, (indicated by annual ring width), and strength-reducing factors such as knots, slope of grain, fissures, reaction wood, fungal and insect damage, and mechanical damage. With the use of grading machines, employed in North America, UK, Australia, New Zealand and Scandinavia since the 1960s, it is possible to determine other characteristics, such as bending modulus of elasticity, that are better correlated with strength properties.

Table B.2 shows the characteristic values for some common strength classes of solid softwood:

- C16 – framing material for timber frame stud walls
- C24 – commonly used as lintel material over openings
- C27 – used to form timber trusses for housing.

Table B.2 Characteristic values for some common strength classes of solid softwood (85)

Property	Symbol	Units	Strength class		
			C16	C24	C27
Characteristic bending strength	$f_{m,k}$	N/mm²	16	24	27
Characteristic tensile strength parallel to the grain	$f_{t,0,k}$		10	14	16
Characteristic tensile strength perpendicular to the grain	$f_{t,90,k}$		0.5	0.5	0.6
Characteristic compressive strength along the grain	$f_{c,0,k}$		17	21	22
Characteristic compressive strength perpendicular to grain	$f_{c,90,k}$		2.2	2.5	2.6
Characteristic shear strength	$f_{v,k}$		1.8	2.5	2.8
Mean value of modulus of elasticity parallel to the grain	$E_{0,mean}$		8000	11000	11500
Fifth percentile value of modulus of elasticity	$E_{0,05}$		5400	7400	7700
Mean value of modulus of elasticity perpendicular to the surface grain	$E_{90,mean}$		270	370	380
Mean value of shear modulus	G_{mean}		500	690	720
Characteristic density	ρ_k[a]	kg/m³	310	350	370
Mean density	ρ_{mean}[b]		370	420	450

[a]Used for calculating the strength of mechanically fastened connections
[b]Used for calculating weight

Bibliography

1 Herbert, G. *Pioneers of Prefabrication.* London: The Johns Hopkins University Press, 1978. ISBN 0 8018 1852 4.

2 Finnimore, B. *House from the factory - System building and the welfare state 1942–74.* 1989. ISBN 1 85489 002 6.

3 Egan, J. *Rethinking Construction.* Department of the Environment, Transport and the Regions. London. 1998. Construction Task Force. ISBN 1 85112 0947.

4 Barker, K. *Modern Methods of Construciton (MMC) for the provision of housing - Technical report covering the barriers to the greater use of modern methods of construction and the mechanisms to overcome them.* s.l. : http://www.scri.salford.ac.uk/MMCFinalReport.pdf, 2006. Barker 33 Cross Industry Group.

5 Market Transformation Programme. *MNMMC03: Modern Methods of Construction - legislation and policy drivers.* Defra. 2008.

6 National Audit Office (NAO). *Using Modern Methods of Construction to Build Homes Quickly and Efficiently.* NAO Information Centre. s.l. : National Audit Office, 2005. DG Ref 5704rf.

7 Falk, A. *Architectural Aspects of Massive Timber - Structural Form and Systems.* Department of Civil and Environmental Engineering, Luleå University of Technology. Luleå: SP External Scientific Publication, 2005. ISSN: 1402-1544.

8 Taylor, F. W. *Principles of Scientific Management.* New York & London : Harper & Bros, 1911.

9 *Knowledge based engineering in construction: the prefabrication timber housing case.* Sandberg, M., Johnsson, H. and Larsson, T. 2008, IT com, Vol. 13.

10 *Construction as a manufacturing process: lessons from the automotive industry.* Crowley, A. 5, 1998, Computers and Structures, Vol. 67, pp. 389–400.

11 *Comparison of cultures in construction and manufacturing industries.* Riley, M. J. and Clare-Brown, D. 3, 2001, Journal of Managment Engineering, Vol. 17, pp. 149–158.

12 *Supply chain management: An empirical study of its impact on performance.* Tan, K. C. et al. 10, 1999, International Journal of Operations & Production Management, Vol. 19.

13 Heikkilä, J. and Suikkari, R. *Log Structures in Finnish Architecture – Continuing the Tradition.* Department of Architecture, University of Oulu. Oulu, Finland : s.n.

14 Nord, T. *Prefabrication Strategies in the Timber Housing Industry - Case Studies from Swedish and Austrian market.* Division of Structural Engineering, Lulea University of Technology. Lulea : s.n., 2008. ISSN 1402 1536.

15 Hook, M. *Timber Volume Element Production - Production and Market Aspects.* Department of Civil and Transport Engineering, Lulea University of Technology. Lulea : s.n., 2005. ISSN 1402 1757.

16 *Re-engineering through pre-assembly: Client expectations and drivers.* Gibb, A. G. F. and Isack, F. 32(2), 2003, Building Research and Information, pp. 146–160.

17 *Construction as a manufacturing process? Similarities and differences between industialized housing and car producion in Japan.* Gann, D. M. 1996, Construction Managment and Economics, Vol. 14, pp. 437–450.

18 *Lean principles in industrialised housing produciton: the need for a cultural change.* Hook, M. and Stehn, L. 2008, Lean Construction Journal.

19 Koskela, L. *An exploration towards production theory and its application to construction.* VTT Technical Research Centre of Finland. 2000. ISBN 951 38 5565 1.

20 Koskela, L. *Application of the new production philosophy to construction.* Stanford University. 1992. Technical Report No. 72.

21 *Matching industrialised timber frame housing needs and enterprise resource planning: A change process.* Bergstrom, M. and Stehn, L. s.l. : Elsevier, 2005, International Journal of Production Economics, pp. 172–184.

22 Ballard, H. G. *The last planner system of production control.* School of Engineering and the Built Environment, Faculty of Engineering, The University of Birmingham. 2000. Doctor of Philosophy Thesis.

23 *A proof for the Queuing Fomula: L = Lamda.W.* Little, J. D. C. 1961, Operations Research, Vol. 9, pp. 383–387.

24 *Exploring the rationales for ERP and SCM integration.* Tarn, J. M., Yen, D. C. and Beaumont, M. 1, s.l.: Industrial Management & Data Systems, 2002, Vol. 102, pp. 26–34.

25 *Enterprise resource planning (ERP) systems: A research agenda.* Al-Mashari, M. 3, 2002, Industrial Management and Data Systems, Vol. 102, pp. 165–170.

26 *Enterprise resource planning business needs and technologies.* Rao, S. 2, 2000, Industrial Management and Data Systems, Vol. 100.

27 *Last Planner - Collaborative production planning, Collaborative programme coordination.* Mossman, A. s.l. : Lean Construction Institute, http://www.leanconstruction.org, 2007.

28 *Last planner: An advance in planning and controlling construction projects. Case studies in the city of Medellin.* Botero, J and Alvarex, M. E. Porto Alegre, Brasil : s.n., 2005. Simposio Bracilerio de Gestao e Economia da Construcao.

29 *Site Implementation and Assessment of Lean Construction Techniques.* Salem, O, et al. s.l. : http://www.leanconstrucitonjournal.org, 2005, Lean Construction Journal, Vol. 2. ISSN 1555 1369.

30 *Analytical Design Planning Technique (ADePT): programming the building design process.* Austin, S. A., Baldwin, A. N. and Waskett, P. R. 1999, Proceedings of the Institution of Civil Engineers, Structures and Buildings, Vol. 134, pp. 111–118.

31 *Requirements and barriers to adoption of last planner computer tools.* Choo, H. J. and Tommelein, I. D. Singapore : http://www.iglc.net/conferences/2001/, 2001. The 9th International Group for Lean Construction Conference.

32 British Standards Institution (BSI). *Design management systems – Part 1: Guide to managing innovation.* s.l. : BSI, 2008. ISBN 978 0 580 58704 7.

33 Goffin, K and Mitchell, R. *Innovation Management - Strategy and Implementation using the Pentathlon Framework.* s.l. : Palgrave MacMillan, 2005. ISBN 978 1 4039 1260 2.

34 Department of Trade and Industry (DTI). *Competing in the global economy: The innovation challenge.* 2003.

35 Chartered Institute of Building. *Code of Practice for Project Management for Construction and Development.* Ascot : Englemere Limited, 2003.

36 *Crane erection of timber trussed rafter roofs.* Hairstans, R., Kermani, A. and Lawson, R. ES2, s.l. : Proceedings of the Institution of Civil Engineers , 2004, Engineering Sustainability, Vol. 157. ISSN 1478 4637.

37 Kazi, A. S. and Eichert, J. *Vision and Strategy of ManuBuild - Open Building Manufacturing.* In Kazi, A. S. et al. Open Building Manufacturing. s.l. : ManuBuild, 2007.

38 Thuesen, C. and Jonsson, C. C. *The long tail and innovation of new construction practices - learning points from two case studies.* In Kazi, A. S. et al. Open Building Manufacturing - Key Technologies, Applications and Industrial Cases. s.l. : ManuBuild, 2009.

39 DTI. *Energy White Paper.* Department of Trade and Industry. Norwich : The Stationery Office, 2003.

40 *Managing knowledge in the context of sustainable construction.* Shelbourn, M. A. et al. 2006, Journal of Information Technology in Construction, Vol. 11, pp. 57–71.

41 *Timber: a growth material for construction.* Dickson, M. 2, 2002, Journal of the Institute of Wood Science, Vol. 16, pp. 78–86.

42 *A Better Quality of Life: A strategy for sustainable development for the UK.* DETR. London : The Stationery Office, 1999.

43 *Engineering buildings for a small planet: Towards construction without depletion.* Dickson, M. London : Institution of Structural Engineers, 21 Feburary 2002, The Structural Engineer, pp. 35–43.

44 DETR. *Building a Better Quality of Life - A Strategy for more Sustainable Construction.* Department of Environment, Transport and the Regions. London : Her Majesty's Stationery Office, 2000.

45 Gaze, C. *Applying the Code for Sustainable Homes on the BRE Innovation Park - Lessons about building fabric.* British Research Establishment. s.l. : BRE Publications, 2008. Information Paper. ISBN 978 1 84806 055 5 Part 1.

46 Gaze, C. and Clift, M. *Applying the Code for Sustainable Homes on the BRE Innovation Park*

- *Lessons learnt about architecture, construction and material sourcing*. British Research Establishment. s.l. : BRE Publications, 2008. Information Paper. ISBN 978 1 84806 065 4 (Part 4).

47 Feist, W., Peper, S. and Gorg, M. *CEPHEUS - Project information No. 36 - Final Report*. s.l. : Passivhaus Institut, 2001.

48 Sustainable Homes Scotland. *Passivhaus - The Backbone of Zero Carbon*. s.l. : Sustainable Homes Scotland, 2009.

49 British Standards Institution (BSI). *BS 5268-2:2002 Structural use of timber - Part 2: Code of practice for permissible stress design, material and workmanship*. 2002. ISBN 0 580 33314 0.

50 British Standards Institution (BSI). *BS EN 1995-1-1:2004: Eurocode 5: Design of timber structures - Part 1.1: General - Common rules and rules for buildings*. 2006. ISBN 0 580 45171 2.

51 British Standards Institution (BSI). *BS EN 1990:2002 Eurocode - Basis of Structural Design*. 2002. ISBN 0 580 40186 3.

52 British Standards Institution (BSI). *BS EN 1991-1-1:2003: Eurocode 1. Actions on structures. General actions. Snow loads*. 2003. ISBN 0 580 423077.

53 IStructE & TRADA. *Manual for the design ot timber building structures to Eurocode 5*. s.l. : The Institution of Structural Engineers, 2007. ISBN 978 0 901297 44 0.

54 *Construction (Design and Management) Regulations*. London : The Stationery Office, 2007. SI 2007/320.

55 Andriamitantsoa, L, D. *Creep*. In Blass, H. J. et al. Timber Engineering STEP 1 - Basis of design, material properties, structural components and joints. 1995, Vol. Centrum Hout.

56 Mettem, C. J., Gordon, J. A and Bedding, B. *Structural Timber Composites*. s.l. : TRADA Technology, 1996. ISBN 1 00510 01 4.

57 *Engineering with the products of the timber industry*. Wells, M. s.l. : The Institution of Structural Engineers, 2006, Structural Engineer.

58 *Design of an 8-storey Residential Tower from KLH Cross Laminated Solid Timber Panels*. Yates, M., Linegar, M. and Dujic, B. Miyazaki : s.n., 2007. 10th World Conference on Timber Engineering.

59 British Research Establishment (BRE). *Multi-storey timber frame buildings a design guide*. s.l. : BRE Press, 2003. BRE Report 454. ISBN 1 860 81605 3.

60 TRADA. *Cross-laminated timber: structural principles*. s.l. : TRADA Technology Ltd, 2009.

61 Chilton, J. C. *History of Timber Structures*. In Blass, H. J. et al. Timber Engineering STEP 2 – Design - Details and structural systems. s.l. : Centrum Hout, 1995.

62 Carmichael, E. N. *Timber Engineering - Practical Designers Guide*. s.l. : E & F. N Spoon, 1984. ISBN 0 419 12700 3.

63 *Load Test on Flitch Beams: Lesson from History*. Desai, S. 4, 2003, The Structural Engineer, Vol. 81, pp. 20–21.

64 *Flitch Beams*. Stern, E. G. and Kumar, V. K. 5, 1973, Forest Products Journal, Vol. 23, pp. 40–47.

65 Larsen, P. and Mettem, C. J. *New Age Flitch - Case study of innovation into practice*. s.l. : TRADA Technology, 2001.

66 Alam, P. *The Reinforcement of Timber for Structural Applications and Repair*. University of Bath. 2004. PhD Thesis.

67 Hairstans, R. *Optimisation of timber platform frame construction*. School of Engineering of the Built Environment, Edinburgh Napier University. 2007. PhD Thesis.

68 *Behaviour of composite timber-concrete floors*. Ahmadi, B. H. and Saka, M. P. 10, 1993, Journal of Structural Engineering, Vol. 119, pp. 3111–3130.

69 *A timber/concrete composite system*. Meierhofer, U. 1993, Structural Engineering Int., pp. 104 - 107.

70 *Experimental verification of new connector for timber-concrete-composite structures*. Sonda, D. Lahti : s.n., 2001. IABSE Conference - Innovative Wooden Structures and Bridges.

71 *Design and testing of a composite timber and concrete floor system*. Persaud, R and Symons, D. s.l. : The Institution of Structural Engineers, 2006, The Structural Engineer.

72 Lukaszewska, E. *Development of prefabricated timber-concrete composite floors*. Luleå : Luleå University of Technology, 2009. ISBN 978 91 86233 85 3.

73 *Performance of Structural Insulated Panels.* Kermani, A. SB1, s.l. : Institution of Civil Engineers, 2005, Stuctures and Buildings, Vol. 159, pp. 13–19.

74 *Racking performance of structural insulated panels.* Kermani, A. and Hairstans, R. 11, 2006, American Society of Civil Engineers (ASCE) Journal of Structural Engineering, Vol. 132, pp. 1806–1812.

75 *Enhancement of the structural properties of glue laminated, homegrown sitka spruce using carbonfibre-reinforcement polymer.* Gilfillan, J, R. 8, 2001, The Structural Engineer, Vol. 79. ISSN 0961 3218.

76 Techniker. *Tall Timber Buildings - The Stadthaus, Hoxton, London: Applications of solid timber construction in multistorey buildings.* London : Techniker, 2010.

77 *Energy-efficient housing - the UK reaches for sustainability.* Banfill, P. F. G and Peacock, A. D. 4, s.l. : Routledge, 2007, Building Research & Information, Vol. 35, pp. 426–436.

78 Department for Communities and Local Government. Cost Analysis of The Code for Sustainable Homes - Final Report. s.l. : Communities and Local Government Publications, 2008.

79 Energy Saving Trust. *Meeting the 10 per cent target for renewable energy in housing - a guide for developers and planners.* London : Energy Saving Trust, 2006.

80 Hoffmeyer, P. Wood as a building material. In Blass, H. J. et al. *Timber Engineering STEP 1 - Basis of design, material properties, structural components and joints.* 1995.

81 Smith, I., Landis, E. and Gong, M. *Fracture and Fatigue in Wood.* s.l. : Wiley, 2003. ISBN 90 5645 001 8.

82 Desch, H. E. and Dinwoodie, J. M. *Timber: its Structural Properties and Utilisations.* 7. s.l. : MacMillan, 1996. ISBN 0 333 25752 9.

83 Dinwoodie, J. M. *Timber its Nature and Behaviour.* s.l. : Van Nostrand Reinhold Company, 2000. ISBN 0356 9403.

84 Young, J. F et al. *The Science and Technology of Civil Engineering Materials.* s.l. : Prentice Hall. ISBN 0 13 659749 1.

85 British Standards Institution (BSI). *BS EN 338:2003 Structural timber - strength classes.* 2003. ISBN 0 580 41845 6.

86 Racher, P. Mechanical timber joints - General. In Blass, H. J. et al. *Timber Engineer Step 1 - Basis of design, material properties, structural components and joints.* 1995.

87 *Towering Timber.* Lowenstein, Oliver. May 2008, The Architects' Journal.

88 *Sustainability into reality: responsible use of timber.* Harris, R. September 2004, The Structural Engineer, pp. 24–27.

89 *21st Century timber engineering - the age of enlightenment for timber design Part 2: Environmental credentials.* Harris, R. London : s.n., December 2005, The Structural Engineer.

90 Eaton, R. A. and Hale, M. D. C. *Wood: Decay, pests and protection.* s.l. : Chapman and Hall, 1993.

91 Ceccotti, A. Timber-concrete composite structures. In Blass, H. J. et al. *Timber Engineering STEP 2 - Design - Details and structural systems.* s.l. : Centrum Hout, 1995.

92 Brunskill, R. W. *Timber Building in Britain.* s.l. : Butler & Tanner Ltd, 1994. ISBN 0 304 36665 X.

93 Anderson, J., Shiers, D. and Steele, K. *The Green Guide to Specification - An environmental profiling system for building materials and components.* s.l. : Wiley-Blackwell, 2009. ISBN 978 1 84806 071 5.

94 *Truss Joist Weyerhaeuser.* Silent Floor & Roof Framing Systems - Technical Guide. 2000.

95 TRADA. *Cross-laminated timber: introduction for specifiers.* s.l. : TRADA Technology Ltd, 2009.

96 Department of Communities and Local Government. *Code for Sustainable Homes - A step change in sustainable home building practice.* London : Communities and Local Government Publications, 2008.

97 Bodig, J. and Jayne, A. B. *Mechanics of Wood and Wood Composites.* Malabar : Krieger Publishing Company, 1993. ISBN 0 89464 777 6.